Evapotranspiration over Spatially Extensive Plant Communities in the Big Cypress National Preserve, Southern Florida, 2007–2010

By W. Barclay Shoemaker, Christian D. Lopez, and Michael J. Duever

Prepared in cooperation with the South Florida Water Management District as part of the U.S. Geological Survey Greater Everglades Priority Ecosystems Science Program

Scientific Investigations Report 2011–5212

U.S. Department of the Interior
U.S. Geological Survey

U.S. Department of the Interior
KEN SALAZAR, Secretary

U.S. Geological Survey
Marcia K. McNutt, Director

U.S. Geological Survey, Reston, Virginia: 2011

For more information on the USGS—the Federal source for science about the Earth, its natural and living resources, natural hazards, and the environment, visit http://www.usgs.gov or call 1–888–ASK–USGS.

For an overview of USGS information products, including maps, imagery, and publications, visit http://www.usgs.gov/pubprod

To order this and other USGS information products, visit http://store.usgs.gov

Suggested citation:
Shoemaker, W.B., and Lopez, C.D., and Duever, Michael, 2011, Evapotranspiration over spatially extensive plant communities in the Big Cypress National Preserve, southern Florida, 2007–2010: U.S. Geological Survey Scientific Investigations Report 2011–5212, 46 p.

Acknowledgments

Steve Krupa of the South Florida Water Management District (SFWMD) provided technical guidance and critical climbing safety features on weather monitoring towers. Patrick Lynch of the SFWMD captured high-resolution photos of the evapotranspiration (ET) sites from a helicopter. Cynthia Gefvert of the SFWMD helped conceptualize the project and locate sites for ET monitoring. Harry, Duke, Joel, Mike, Ozzie, and Papo of Kirms Communications, Inc., and Sphere Communications, Inc., constructed the ET towers. Damon Doumlele of the National Park Service helped draft the "Environmental Assessment and Finding of No Significant Impact" documents. Anonymous firefighters are acknowledged for digging small trenches around the perimeter of the Marsh ET site during a wildfire, preventing fire damage to the instrumentation. Brian Bovard from Florida Gulf Coast University installed sap flow sensors to quantify transpiration from the forested ET sites. Troy Bernier from H_2O Resource and Jason McCobb provided field assistance. Mike Barry, Maureen Bonness, and Jean McCollom are acknowledged for detailed characterizations of the plant communities at the ET sites.

The following USGS employees are acknowledged for their assistance. David Sumner, Amy Swancar, and Raymond Dupuis provided field and technical guidance and assisted with tower climbing. Trey Grubbs reviewed and improved spreadsheets used for ET calculations. Rhonda Howard provided editorial review comments. Amy Swancar, David Sumner, Dorothy Payne and Robert Renken provided peer-review and technical comments that improved the report quality.

Contents

Figures

Tables

Conversion Factors and Abbreviations

Multiply	By	To obtain
Length		
millimeter (mm)	0.03937	inch (in)
centimeter (cm)	0.3937	inch (in)
meter (m)	3.281	foot (ft)
kilometer (km)	0.6214	mile (mi)
Area		
hectare (ha)	2.471	acre
square meter (m^2)	10.76	square foot (ft^2)
Flux		
cubic meter (m^3)	35.31	cubic foot (ft^3)
meter per second (m/s)	3.281	foot per second (ft/s)
millimeter per day (mm/d)	0.03937	inch per day (in/d)
millimeter per year (mm/yr)	0.03937	inch per year (in/yr)
Flow rate		
cubic meter per second (m^3/s)	35.31	cubic foot per second (ft^3/s)
Mass		
gram (g)	0.03527	ounce, avoirdupois (oz)
Pressure		
kilopascal (kPa)	0.2961	inch of mercury at 60°F (in Hg)
kilopascal (kPa)	0.1450	pound-force per square inch (lbf/in^2)
Energy flux density		
watt per square meter (W/m^2)	0.001433	calorie per square centimeter per minute (cal/cm^2/min)
Energy		
joule (J)	0.2388	calorie(cal)

List of Variables and Symbols

A Changes in heat energy storage in the air column beneath the eddy covariance instrumentation, in watts per square meter

A_e Energy available for latent and sensible heat, in watts per square meter

α Priestley-Taylor calibration coefficient, unitless

α_2 Fraction of an air-temperature change that eventually causes a water-temperature change, unitless

B Ratio of sensible- to latent-heat flux (Bowen ratio), unitless

λE Latent heat flux, in watts per square meter

λE_{cor} Corrected latent-heat flux, in watts per square meter

C_A Specific heat capacity of moist air, in joules per gram degree Celsius

C_p Specific heat capacity of air, in joules per gram degree Celsius

C_s Specific heat capacity of soil, in joules per gram degree Celsius

C_{sd} Specific heat capacity of dry soil, in joules per gram degree Celsius

C_w Specific heat capacity of water, in joules per gram degree Celsius

d Soil layer thickness, in centimeters

d_{ec} Distance from the eddy covariance instrumentation to land surface, in meters

d_w Water depth, in meters

dLE Eddy covariance instrumentation

D_{i-j} Water distance above land surface, in meters, at time i within the surface water's thermal memory j

D_b Dry-soil bulk density, in grams per cubic centimeter

$\overline{ET}_{min}$ Mean evapotranspiration (ET) over 3 months with the least amount of ET

$\overline{ET}_{max}$ Mean evapotranspiration (ET) over 3 months with the greatest amount of ET

F Krypton hygrometer correction factor equal to 0.229 gram degree Celsius per joule

G Soil heat flux at land surface, in watts per square meter

H Sensible heat flux, in watts per square meter

H_{cor} Corrected sensible-heat flux, in watts per square meter

Δh Change in water level, in millimeters

$IMEM$ Thermal memory of surface water, in days

K_{ex} Thermal exchange coefficient, in meters per second

K_o Extinction coefficient of hygrometer for oxygen, in cubic meters per gram per centimeter

K_w Extinction coefficient of hygrometer for water, in cubic meters per gram per centimeter

ΔLE Changes in latent-heat storage in the air column beneath the eddy covariance instrumentation, in watts per square meter

ρ'_v	Instantaneous variations of water-vapor density from a mean value, in grams per cubic meter
ρ_a	Air density, in grams per cubic meter
$\Delta\rho_v$	Change in vapor density, in grams per cubic meter
P	Precipitation, in millimeters
P_a	Atmospheric pressure, in pascals
Q_{in}	Surface-water inflow, in cubic meters per second
Q_{out}	Groundwater outflow, in cubic meters per second
R_d	Gas constant for dry air, in joules per gram degree Celsius
R_n	Net radiation, in watts per square meter
S	Changes in heat energy storage in the soil, in watts per square meter
S_y	Specific yield, unitless
S_i	Seasonality index
T_a	Air temperature, in degrees Celsius
$\Delta T_{a_{i-j}}$	Air temperature change, in degrees Celsius, at time i within the surface water's thermal memory j
T_s	Instantaneous variations of sonic air temperature from a mean value, in degrees Celsius
Δt	Change in time, in seconds
ΔT_s	Change in soil temperature, in degrees Celsius
ΔT_w	Change in water temperature, in degrees Celsius
W	Changes in heat-energy storage in surface water, in watts per square meter
w'	Instantaneous variations of vertical wind speed from a mean value, in meters per second
X_w	Mass fraction of water in the soil, in grams water per grams dry soil
γ	Psychrometric constant, in kilopascals per degree Celsius
Δ	Slope of the saturated vapor pressure curve, in kilopascals per degree Celsius

Evapotranspiration over Spatially Extensive Plant Communities in the Big Cypress National Preserve, Southern Florida, 2007–2010

By W. Barclay Shoemaker[1], Christian D. Lopez[1], and Michael J. Duever[2]

Abstract

Evapotranspiration (ET) was quantified over plant communities within the Big Cypress National Preserve (BCNP) using the eddy covariance method for a period of 3 years from October 2007 to September 2010. Plant communities selected for study included Pine Upland, Wet Prairie, Marsh, Cypress Swamp, and Dwarf Cypress. These plant communities are spatially extensive in southern Florida, and thus, the ET measurements described herein can be applied to other humid subtropical locations such as the Everglades.

The 3-year mean annual ET was about 1,000, 1,050, 1,100, 930, and 900 mm (millimeters) at the Dwarf Cypress, Wet Prairie, Cypress Swamp, Pine Upland, and Marsh sites, respectively. Spatial differences in annual ET were considerable due to the recovery of the Marsh site from extensive forest fire and drought conditions. Temporal variability in annual ET was relatively small at sites that were well watered (Dwarf Cypress, Wet Prairie, Cypress Swamp, Pine Upland) over the 3-year study. In other words, locations that were well watered appeared to have similar annual ET rates.

Diurnal water-level variability was observed in response to ET and was less at flooded sites than at dry sites. For example, surface-water levels declined about 1.5 mm in response to ET at the flooded Cypress Swamp site during July 18-22, 2008 and declined about 10 mm in response to ET at the flooded Dwarf Cypress site from April 18-27, 2008. Specific yield was computed using ET estimates and diurnal water-level variability measured at the dry Pine Upland site as a check on the accuracy of the eddy covariance method. Water levels repeatedly dropped about 15 mm on average in response to ET at the Pine Upland site from April 27 to May 4, 2008. ET was about 3 mm on each of these days, resulting in a reasonable estimate for specific yield of 0.2 for the sandy soils at the Pine Upland site.

Monthly ET estimates exhibited seasonal variation. ET was generally greatest between March to October when solar radiation was relatively large, and least from November to February when solar radiation was small. Monthly ET was greatest in the spring and summer at the Cypress Swamp site, reaching rates as large as 140 mm. The large ET rates at this site coincide with the most active period of cypress growth during late spring and early summer. Cypress trees begin to senesce in late summer reducing transpiration.

Net radiation and available energy explained most of the variability in ET observed at all five sites. Mean annual and monthly net radiation varied among the sites in response to cloud cover and the albedo of the land surface and plant community. Net radiation was greatest at the Cypress Swamp site, averaging about 130 W/m^2 (watts per square meter) during the 3-year study. Net radiation was generally less at the Dwarf Cypress site, averaging about 115 W/m^2 over 3 years. The Dwarf Cypress site apparently has the largest albedo, which likely is due to the sparse canopy and a highly reflective, calcareous, periphyton-covered land surface. Furthermore, mean annual net radiation was least in the first year of the study, which likely was due to greater cloud cover during a relatively wet year. In contrast, net radiation was greatest in the second year of the study, which likely was due to less cloud cover during a relatively dry year.

Introduction

Evapotranspiration (ET) is the rate of transport of water vapor from the earth's surface into the atmosphere and represents a composite flux of surface water directly evaporated by solar energy, and subsurface water transpired by plants. Factors limiting this major component of the hydrologic cycle include available energy, available water, and vapor-transport resistance offered by the atmosphere and vegetation. ET is very important in the hydrologic cycle of southern Florida, transporting as much as 80 to more than 100 percent of rainfall back into the atmosphere as water vapor.

Several researchers (Bidlake and others, 1996; Knowles, 1996; Sumner, 1996, 2001; and German, 2000) have successfully applied the eddy covariance method (Dyer, 1961; Tanner and Greene, 1989) to directly measure ET in Florida. The

---

[1] U.S Geological Survey

[2] Natural Ecosystems L.L.C.

eddy covariance method is a micrometeorological approach currently in use within the Florida ET network (Sumner, 2001), as well as national (AMERIFLUX) and international (FLUXNET) climate-monitoring networks. The eddy covariance method measures both latent- and sensible-heat fluxes transported by turbulent wind eddies in the air. Latent-heat flux (λE) is the energy equivalent of ET. Sensible-heat flux (H) is the heat energy removed from the earth's surface due to processes such as convection. The eddy covariance method provides several advantages relative to other measurement methods, including more areal integration, less site disruption within environmentally sensitive areas, elimination of the

need to estimate other water-budget terms, and finer temporal resolution (30-minutes).

In 2005, the U.S. Geological Survey (USGS) and South Florida Water Management District (SFWMD) began a cooperative study to measure ET in southern Florida using the eddy covariance method. Several spatially extensive plant communities were studied including Dwarf Cypress, Cypress Swamp, Pine-Upland, Wet Prairie, and Marsh as mapped by Duever and others (1986) within the Big Cypress National Preserve (BCNP) (fig. 1). These plant communities are common in humid subtropical south Florida. Thus, the ET measurements described herein can be applied to other humid subtropical locations such as the Everglades.

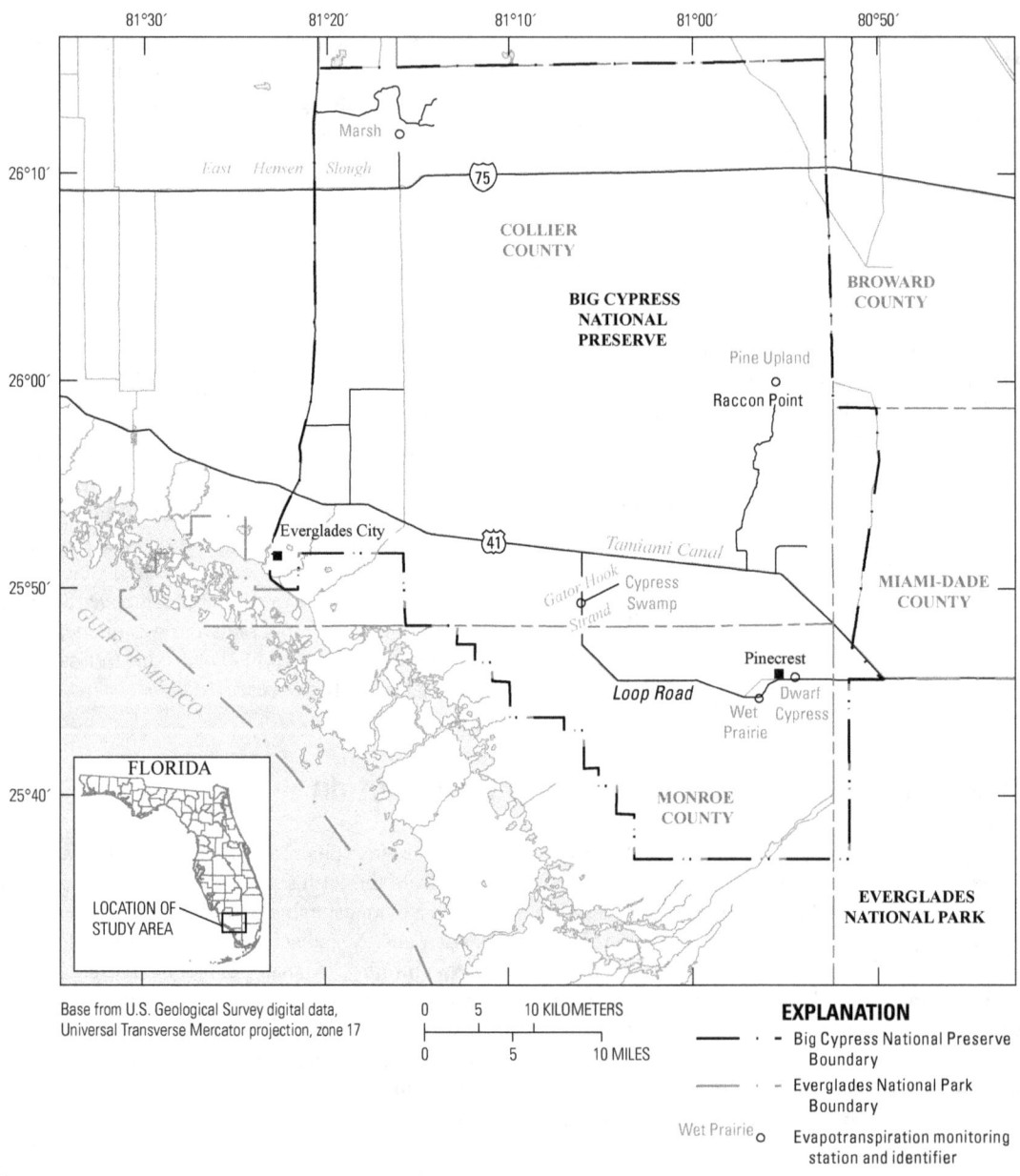

Figure 1.   Big Cypress National Preserve study area and evapotranspiration monitoring stations in southern Florida.

During the study period, the area experienced cold fronts, heat waves, tropical storms, hurricanes, floods, droughts, and forest fires. In January 2010, for example, a severe cold front moved through southern Florida with night-time temperatures reaching zero degrees Celsius for several days. Furthermore, Tropical Storms Barry and Faye traveled with minimal damage through the study area on June 1, 2007 and August 18, 2008, respectively, and Hurricane Gustav passed through southern Florida on September 1, 2008.

Purpose and Scope

This report quantifies ET over spatially extensive plant communities within the BCNP over 3 years. Five ET monitoring stations were established in representative Pine Upland, Wet Prairie, Marsh, Cypress Swamp, and Dwarf Cypress plant communities. The eddy covariance method was applied to measure high-resolution (30-minute) values of latent-heat flux, the energy-equivalent of ET. A modified Priestley-Taylor model estimated latent-heat flux during periods when measured data were unavailable. The high-resolution data were upscaled into mean monthly and annual ET values. Hydrologic and climatic conditions that explain ET variability were identified, and the response of ET to events such as flooding, drought and cold fronts was examined.

Description of Study Area

The study area is the BCNP in southern Florida (fig. 1), which is the first National Preserve in the National Park System. The Preserve consists of 570,000 acres of primarily cypress, pine wet-prairie, marsh, and hardwood hammock plant communities in Miami-Dade, Collier and Monroe Counties. The BNCP was established by an Act of Congress in 1974 in response to political advocacy by various stakeholders interested in protecting an ecologically sensitive area from encroachment by development. The National Preserve designation was chosen in order to permit continuation of anthropogenic activities, including oil and gas production, hunting, grazing, and off-road vehicle use, which are normally not allowed in national parks.

The BCNP's wetlands provide a wide range of ecosystem services, including floodwater protection, erosion buffering, substrate stabilization, sediment trapping, and water filtration for extensive coastal estuaries. The wetlands also serve as habitat for numerous species of birds, mammals, reptiles, amphibians, fish, and insects such as mosquitoes. The extensive, shallow, gray limestone aquifer (Reese and Cunningham, 2000) underlies the BCNP. The gray limestone aquifer is composed of porous limestone that is about 15 to 30 m thick on the western boundary of the BCNP, and generally diminishes in thickness to the east (Shoemaker, 1998; Reese and Cunningham, 2000). The gray limestone aquifer is the primary source of fresh drinking water in Collier County.

Climate in the BNCP is humid subtropical characterized by a hot, humid wet season and a mild dry season. The wet season generally begins between mid-May and early June and continues into September or early October. Daily thunderstorms or local showers can occur during the wet season. Tropical storms and hurricanes occasionally pass through the area creating heavy rainfall. During the dry season, precipitation is usually associated with large frontal systems.

Plant Communities and Vegetation Classification

The BNCP hosts a variety of plant communities, including Pine Upland, Wet Prairie, Marsh, Hardwood Hammock, Cypress Swamp, and Dwarf Cypress (McPherson 1973; Duever and others 1986). Marsh and Cypress Swamp typically occupy low elevations, Wet Prairie occupies middle elevations, and Pine Upland and Hardwood Hammock occupy high elevations.

The distribution of plant communities in the BCNP is controlled by topography, hydrologic and fire regimes, and soil conditions (Duever and others, 1986). The Wet Prairie site is classified as muhly freshwater prairie and supports a low (less than 1 m) herbaceous plant community dominated by muhly grass (*Muhlenbergia* capillaries), sawgrass (*Cladium jamiacense*), and a large number of other species of grasses, sedges, and forbs. Periphyton is abundant in this open (20 to 40 percent cover) community. The substrate is marl over a topographically irregular limestone bedrock surface that is generally close to, or in places, at land surface.

The Dwarf Cypress site is classified as scrub cypress-sawgrass and is dominated by scattered stunted cypress, *Taxodium distichum*, and scattered (5 to 10 percent cover) sawgrass less than 1 m high. Small-to medium-sized stands of cypress are present, and periphyton is abundant. The substrate is shallow marl overlaying topographically irregular limestone bedrock. Wet conditions at this site resulted in deeper marl sediments than at the Wet Prairie site, and little or no exposure of bedrock at land surface.

The Cypress Swamp site is classified as a swamp forest and supports a tall dense cypress forest with a subcanopy of mixed hardwoods. The varieties include holly (*Ilex cassine*), swamp bay (*Persea palustris*), maple (*Acer rubrum*), an open-to-dense shrub layer with coco plum (*Chrysobalanus icaco*), myrsine (*Rapanea punctata*), wax myrtle (*Myrica cerifera*), an open-to-dense ground cover of swamp fern (*Blechnum serrulatum*), and a variety of grasses, sedges, and forbs. The substrate is primarily topographically irregular limestone bedrock with organic accumulations in depressions in the rock.

The Marsh site is classified as a mix of *graminoid* and broadleaf emergent freshwater marsh vegetation. The *graminoid* marsh is an open community (20 to 30 percent cover) on slightly higher sites dominated by sand cordgrass, *Spartina bakeri*, and a variety of other grasses, sedges, and forbs. The sand cordgrass is generally less than 2 m high, whereas the other components are generally only about 1 m high. The broadleaf community is much denser (80 to 90 percent cover), with a vegetation height of about 1.5 m, and dominated by pickerel weed (*Pontederia cordata*), maidencane (*Panicum hemitomom*), arrowhead (*Sagittaria lancifolia*), and small patches of willow (*Salix caroliniana*) less than 3.5 m high in the deepest places. Substrates in the graminoid marsh are deep sands that grade into organic sands with increasing water depth and finally deep organic muck in the broadleaf and willow communities.

The Pine Upland site, classified as a mixed lowland pine site, is located in an extensive open-canopy pine forest with numerous small- to medium-sized cypress domes. The canopy is dominated by slash pine (*Pinus elliottii*); patches of saw palmetto (*Serenoa repens*); small trees and shrubs including holly, swamp bay, myrsine, and wax myrtle; and scattered sabal palms (*Sabal palmetto*). The ground cover is a diverse mix of short (less than 1 m) grasses, sedges, and forbs that are scattered in open-to-dense patches around the site. The open character of the site indicates the regular occurrence of fire, probably on the order of +5 or more years. The larger cypress domes have a dense canopy of cypress, but open subcanopy and shrub strata, probably because of the frequent occurrence of fire. The ground cover also is sparse in the center of the domes, but becomes somewhat more abundant toward the dome edges. Forbs are more dominant within the deeper center of the domes, and grasses and sedges become more dominant toward the shallower edges of the domes. Substrates are primarily limestone bedrock in the Pine Upland, with sandy marl in the shallow depressions. The cypress dome has a shallow organic substrate in the deeper areas.

Previous Studies

Previous ET studies in Florida include Abtew and Obeysekera (1995), who compared the abilities of Penman combination methods and the Priestly-Taylor equation to reproduce wetland ET values obtained with a lysimeter. Abtew (1996a; 1996b) developed and tested three simple solar radiation equations for computing ET in southern Florida. German (2000) completed a benchmark study using the Bowen-ratio method that estimated ET ranging from about 1,000 to 1,400 mm/yr at nine sites in the Everglades. Shoemaker and Sumner (2005) derived a new approach for computing changes in heat energy stored in wetland surface water, a considerable component of subdaily and daily surface-energy budgets. Shoemaker and Sumner (2006) also developed numerous corrections for estimating actual wetland ET derived from potential ET. Price and others (2007) have considered the uncertainty associated with evaporation estimates due to measurement error, and estimated a mean annual evaporation of about 1,700 mm/yr from Florida Bay over a 33-year period. Shoemaker and others (2008) tested eight representations of ET in a simplified wetland

hydrologic model. In that study, an ET representation that included extinction depth was found to potentially underestimate the annual volume of water available for groundwater recharge and surface-water runoff to coastal estuaries surrounding Everglades National Park by 2.3 billion m[3]. For comparison, this underestimation is about two thirds of the mean volume of water in Lake Okeechobee (3.8 billion m[3]), the largest lake in Florida, and clearly demonstrates unbiased estimates of ET are necessary for reliably calculating wetland water budgets.

In central Florida, Bidlake and others (1996) measured ET over native vegetation using both energy-budget Bowen ratio and eddy covariance methods. Sumner (2001) estimated ET at about 900 mm in 1998 and 1,000 mm in 1999 over a cypress and pine forest that was subjected to logging and natural fires in east-central Florida. Sumner (2006) studied the adequacy of selected ET approximations for hydrologic simulation. Swancar and others (2000) measured evaporation that exceeded rainfall within Lake Star in central Florida. At that lake, rainfall was about 1,300 and 1,370 mm/yr for these 2 years of study. Lake evaporation was 1,450 and 1,422 mm/yr for the same 2 years, making available water (rainfall minus evaporation) negative during both years.

Satellites also can be used to remotely estimate ET. Specifically, satellite images of the Earth can be used to develop estimates of solar radiation that are critical to ET calculations (Jacobs and others, 2008). Satellite-derived ET estimates offer greater spatial coverage than ground-based ET networks, but are less reliable under cloudy conditions; field-based ET measurements are needed as a reality check for satellite ET estimates. Maps of ET were compiled by Islam and others (2002), who estimated a coefficient in the Priestly-Taylor equation with satellites. Jacobs and others (2008) used satellite imagery to provide gridded estimates of solar radiation, net radiation, potential ET, and reference ET for all of Florida.

Methods for Measurement of Evapotranspiration

The eddy covariance energy-budget method (Dyer, 1961; Tanner and Greene, 1989; Bidlake and others, 1996) was used to measure ET for this study. This method is generally considered a direct and accurate technique for measuring ET (Bidlake and others, 1996; Sumner 2001), and therefore, was selected for this study. Surface-energy budgets, required for the eddy covariance energy-budget method, are introduced and discussed. Also documented are the factors considered during site selection including tower "foot prints" or source-measurement areas. Finally, quality assurance and quality control procedures are summarized, including estimation of missing records according to procedures outlined by Sumner (1996; 2001) and German (2000).

Conceptualization of the Surface-Energy Budget

The surface-energy budget (fig. 2, eq. 1) governs the energy available for ET. Net radiation (R_n) is the difference between incoming radiation (shortwave solar radiation and long-wave atmospheric radiation) and outgoing radiation (reflected shortwave and upwelling long-wave radiation). Net radiation is the source for changes in heat energy stored in the canopy, soil (S) and surface water (W). Changes in latent heat stored in the air column beneath the eddy covariance instrumentation (dLE) and changes in heat energy within the air column (A) beneath the eddy covariance instrumentation are additional storage terms that can be approximated with available data. Thermal energy passing through the soil is commonly called the soil-heat flux (G). The energy available (A_e) for latent and sensible heat is defined as the difference between net radiation and the sum of the storage-change terms and the soil-heat flux (eq. 1). Assuming the net horizontal advection of energy and changes in heat energy stored in the canopy are both negligible; the simplified surface-energy budget equation takes the form:

$$R_n - (S + W + G + A + dLE) = \lambda E + H \qquad (1)$$

where the units of each energy flux are in watts per square meter. Latent-heat flux (λE) is defined as energy removed from the surface in the liquid-to-vapor phase change of water. Sensible-heat flux (H) is the heat energy removed from the surface due to a temperature gradient between the canopy and the air. Both latent- and sensible-heat fluxes are directly measured with the eddy covariance instrumentation. Likewise, net radiation (R_n) and the soil heat flux (G) are directly measured with net radiometers and soil heat flux probes, respectively. The remaining heat-energy storage terms were computed with ancillary data as described below.

Changes in heat energy stored in the shallow (0-20 cm) soil (S, eq. 1) were computed using measurements of soil-temperature change in the soil column and soil-moisture content, when available. The soil heat-flux plate was buried about 20 cm below land surface. The change in soil-heat storage was computed as (Campbell Scientific, Inc., 1990; German, 2000):

$$S = \frac{10,000 \Delta T_s C_s d}{\Delta t} \qquad (2)$$

where 10,000 is a conversion factor; ΔT_s is soil temperature change, in degrees Celsius; C_s is the volumetric heat capacity of the soil, in joules per gram degree Celsius; d is soil layer thickness (20 cm); and Δt is the time interval (1,800 seconds). When measurements of soil moisture were available, the soil heat capacity (C_s) was estimated from the relation:

$$C_s = D_b (C_{sd} + C_w X_w) \qquad (3)$$

where D_b is the dry-soil bulk density (assumed to be 1.5 grams per cubic centimeter); C_{sd} is the specific heat capacity of the dry soil (assumed to be 0.840 joule per gram degree Celsius);

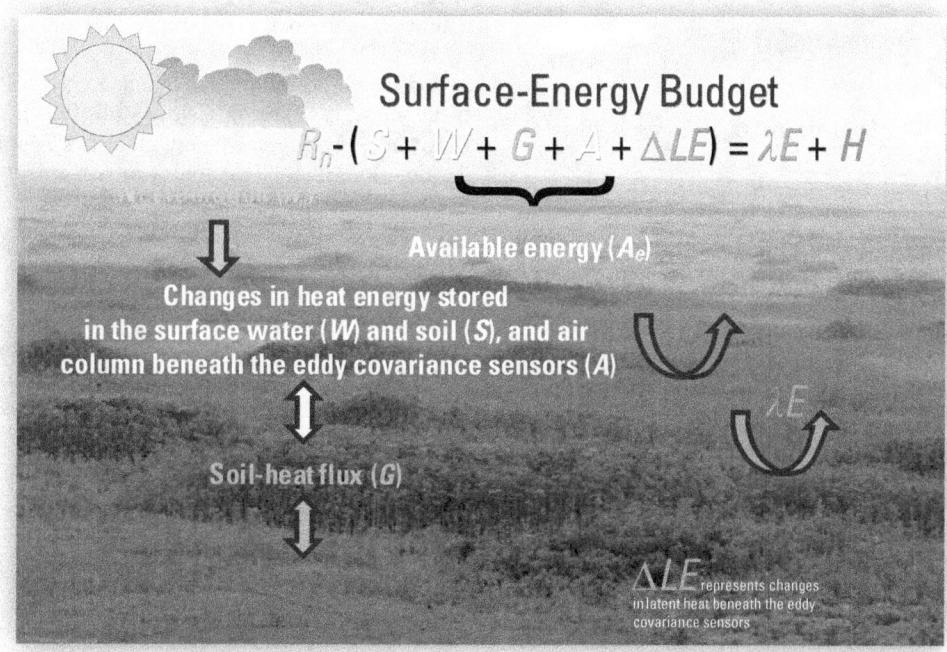

Figure 2. Conceptualization of the surface-energy budget. Photograph by Patrick Lynch, South Florida Water Management District.

C_w is the specific heat capacity of water (4.190 joules per gram degree Celsius); and X_w is the mass fraction of water in the soil (grams water per grams dry soil).

Water temperatures at land surface and about 152 mm (6 in.) above land surface were measured to quantify changes in heat energy stored in surface water. These data were unusable because the sensors were mounted on steel poles that created spikes in water temperature that were related more to the steel temperature than to the water temperature. Therefore, water temperature changes in the surface water were estimated through convolution of air-temperature changes with a regression-defined transfer function, as outlined by Shoemaker and others (2005). In summary, 30-minute water-temperature changes were estimated as:

$$\Delta T_{w_i} = \sum_{j=1}^{IMEM} \frac{K_{ex}}{D_{i-j}} e^{\frac{K_{ex}}{D_{i-j}}} \alpha_2 \Delta T_{a_{i-j}}$$

(4)

where ΔT_w is the water-temperature change in degrees Celsius; i is the integer time step for computing 30-minute changes in surface-water temperature; j is the integer time step discretizing the surface water's thermal memory ($IMEM$) equal to 1 day; K_{ex} is a thermal exchange coefficient, in meters per second; D_{i-j} is the depth of water above land surface, in meters; α_2 is the fraction of an air-temperature change that eventually causes a water-temperature change; and $\Delta T_{a_{i-j}}$ is the air-temperature change in degrees Celsius. The variables K_{ex} and α_2 are regression defined to compute changes in water temperature measured in the Everglades by German (2000). The mean of K_{ex} and α_2 values presented in Shoemaker and others (2005, table 2) were used in equation 4, equal to 2.021 m/s and 0.607 respectively.

Changes in heat energy stored in the surface water (W, equation 1) were computed as presented in German (2000):

$$W = \frac{304,800 d_w \Delta T_w C_w}{\Delta t}$$

(5)

where 304,800 is a conversion factor; d_w is the water depth, in meters; ΔT_w is the convolution-computed water-temperature change, in degrees Celsius; C_w is the heat capacity of water (4.19 joules per gram degree Celsius); and Δt is the time interval (1,800 seconds).

Changes in latent heat stored in the canopy beneath the eddy covariance instrumentation (dLE) were computed as:

$$dLE = \frac{\Delta \rho_v d_{ec} \lambda}{\Delta t}$$

(6)

where $\Delta \rho_v$ is the vapor density change, in grams per cubic meter, computed as a function of air temperature and vapor pressure over the 30-minute averaging period Δt; and d_{ec} is the distance from the eddy covariance instrumentation to land surface, in meters. The change in heat energy within the air

column beneath the eddy covariance instrumentation (A) was estimated in a similar manner as:

$$A = \frac{C_A \Delta T_a}{\Delta t} d_{ec} \rho_a$$

(7)

where C_A is the specific heat of moist air, in joules per gram degree Celsius; ΔT_a is the air temperature change, in degrees Celsius; and ρ_a is the air density, in grams per cubic meter.

Eddy Covariance Method

The eddy covariance method is a one-dimensional approach for measuring latent- and sensible-heat fluxes within the atmospheric surface layer (Campbell and Norman, 1998). Latent- and sensible-heat fluxes are approximated as:

$$\lambda E = \overline{w' \rho'_v}$$

(8)

$$H = \rho_a C_p \overline{w' T_s'}$$

(9)

where λE is the latent-heat flux, in watts per square meter; w' represents instantaneous variations of vertical wind speed from a mean value, in meters per second; ρ'_v represents instantaneous variations of water-vapor density from a mean value, in grams per cubic meter; H is sensible heat in watts per square meter; ρ_a is the air density, in grams per cubic meter, estimated as a function of air temperature, total air pressure, and vapor pressure (Monteith and Unsworth, 1990); C_p is the specific heat capacity of air, in joules per gram degree Celsius, estimated as a function of temperature and relative humidity (Stull, 1988); and T_s' represents instantaneous variations of sonic air temperature (Kaimal and Gaynor, 1991) from a mean value, in degrees Celsius (Stull, 1998). The overbars ($\overline{w' p'_v}, \overline{w' T_s'}$) represent averaging over a 30-minute period. Assumptions include the source measurement area is roughly horizontal and the net lateral advection of water vapor is negligible. Further assumptions can be found in Campbell and Norman (1998).

Numerous errors occur in eddy flux measurements due to assumption violations, instrumentation failures, and physical phenomena. These errors may exceed the initial flux estimate and therefore are not trivial (Burba and Anderson 2007). Latent- and sensible-heat fluxes were corrected according to methods outlined by Sumner (1996; 2001). In summary, latent-heat flux was corrected using the following equation:

$$\lambda E = \lambda \left[\overline{(w' \rho'_v)} + \frac{\rho_v H}{\rho_a C_p (T_a + 273.15)} + \frac{F K_o H}{K_w (T_a + 273.15)} \right]$$

(10)

where T_a is air temperature, in degrees Celsius; F is a factor that accounts for molecular weights of air and atmospheric abundance of oxygen equal to 0.229 gram degree Celsius per joule; K_o is the hygrometer extinction coefficient for oxygen estimated as 0.0045 cubic meter per gram per centimeter (Tanner and others, 1993); K_w is the hygrometer extinction coefficient for water equal to the manufacturer-calibrated value in cubic meters per gram per centimeter. The overbars and primes indicate means over the averaging period and deviations from the means, respectively. The second and third terms of the right side of equation 10 account for temperature-induced fluctuations in air density (Webb and others, 1980) and for the sensitivity of the hygrometer to oxygen (Tanner and Greene, 1989), respectively.

Sensible heat was corrected for the effects of wind blowing normal to the sonic acoustic path (E. Swiatek, Campbell Scientific, Inc., written commun., 1998). Furthermore, Schotanus and others (1983) have related the sonic sensible heat to the true sensible heat as follows:

$$H = \rho_a C_p \frac{\overline{T_a}}{\overline{T_s}} \left(\overline{w'T_s'} - \frac{0.51 R_d \overline{(T_a + 273.15)^2 \, w'\rho'v}}{P_a} \right) \quad (11)$$

where T_a is air temperature, in degrees Celsius; R_d is the gas constant for dry air (0.28704 joule per gram degree Celsius); and P_a is atmospheric pressure, in pascals. Turbulent fluxes were corrected for mis-leveling of the sonic anemometer, as outlined by Tanner and Thurtell (1969), Baldocchi and others (1988), and Sumner (2001).

Several researchers (Moore, 1976; Goulden and others, 1996; German, 2000) have noted the eddy covariance method performs best in windy conditions (relatively high friction velocity, u*). German (2000) noted that at u* greater than 0.3 m/s, less discrepancy existed between measured available energy and measured turbulent fluxes in the Everglades. Thus, turbulent fluxes were culled and gap filled when u* was less than 0.3 m/s.

Previous investigators (Lee and Black, 1993; Bidlake and others, 1996; Sumner, 1996; and German, 2000) describe a recurring problem with the eddy covariance method. Specifically, the sum of the measured latent- and sensible-heat fluxes is generally less than the measured available energy. Foken (2008) explained this discrepancy with low-frequency (large-scale) eddies unmeasured by the chosen averaging period (usually 30 minutes) of equations 8 to 11. Measured 30-minute latent- and sensible-heat fluxes were corrected to account for low-frequency (large scale) eddies by assuming the ratio of turbulent fluxes (Bowen ratio) was adequately measured (Moore, 1976), and partitioning the residual available energy by the Bowen ratio (Bowen, 1926), where the Bowen ratio (B) was calculated as:

$$B = \frac{H}{\lambda E} \quad (12)$$

and the residual available energy $(A_e-[\lambda E+H])$ was partitioned into corrected latent and sensible heat using the following equations:

$$A_e = H_{cor} + \lambda E_{cor} = \lambda E_{cor}(1+B) \quad (13)$$

$$\lambda E_{cor} = \frac{A_e}{1+B} \quad (14)$$

$$H_{cor} = A_e - \lambda E_{cor} \quad (15)$$

where λE_{cor} is the corrected latent-heat flux, in watts per square meter; and H_{cor} is the corrected sensible-heat flux, in watts per square meter, reported herein.

Source Area and Site Selection

The source area for turbulent flux measurements is the upwind land-surface area contributing water vapor and heat to the eddy covariance sensors. Instrumentation was mounted on towers at heights that seemed likely to capture only Cypress Swamp, Dwarf Cypress, Pine Upland, Marsh, and Wet Prairie ET rates (fig. 3), given source-area calculations derived from Scheupp and others (1990). The source area was defined as the radial distance surrounding a tower that likely contributes greater than 90 percent of the total flux measurement.

Gator Hook Strand was chosen for the Cypress Swamp ET site (table 1 and fig. 1). Gator Hook Strand extends about 1,600 m from northwest to southeast with 18- to 25-m tall cypress trees. Tree height guided estimation of tower and instrumentation mounting heights above the cypress canopy. According to Scheupp and others (1990), flux sensors placed about 12 m above a 24-m canopy height create a radial source area extending about 800 m from the tower, assuming a roughness length of 2.4 m (about 10 percent of the canopy height), a displacement height of 15.6 m (about 65 percent of the canopy height), and a "mildly unstable" atmosphere with an Obukhov length equal to -10 (fig. 4). A "mildly unstable" atmosphere allows vertical motion of air parcels with water vapor. At night, the atmosphere is likely stable (vertical motion is limited) and the footprint extends much farther than 800 m. During the day, however, when ET is greatest, the atmosphere could be "mildly unstable" to "very unstable" with an Obukhov length equal to -10 and -1, respectively. Radial source areas would be substantially smaller (about 400 m) when the atmosphere is "very unstable." Thus, the eddy covariance instrumentation was mounted about 12 m above the mean canopy height on a 38-m tower (fig. 4), so the source area was large cypress trees under "mildly unstable" to "very unstable" atmospheric conditions.

An area north of Loop Road was chosen for the Dwarf Cypress ET site (table 1 and fig. 1). Relatively small cypress trees (4-10 m high) with a sawgrass understory extend for many kilometers in every direction. Tree height guided estimation of tower and instrumentation heights above the small cypress canopy. According to Scheupp and others (1990), sensors placed about 6 m above a 10-m canopy create a radial source area extending about 400 m from the tower, assuming a roughness length of 1m (about 10 percent of the canopy height), a displacement height of 6.9 m (about 65 percent

A. Cypress Swamp

Figure 3.   Photographs of the evapotranspiration monitoring stations. Photograph by Patrick Lynch, South Florida Water Management District.—Continued

B. Dwarf Cypress

Figure 3.   Photographs of the evapotranspiration monitoring stations. Photograph by Patrick Lynch, South Florida Water Management District.—Continued

C. Pine Upland

Figure 3. Photographs of the evapotranspiration monitoring stations. Photograph by Patrick Lynch, South Florida Water Management District.—Continued

D. Marsh

Figure 3.   Photographs of the evapotranspiration monitoring stations. Photograph by Patrick Lynch, South Florida Water Management District.—Continued

E. Wet Prairie

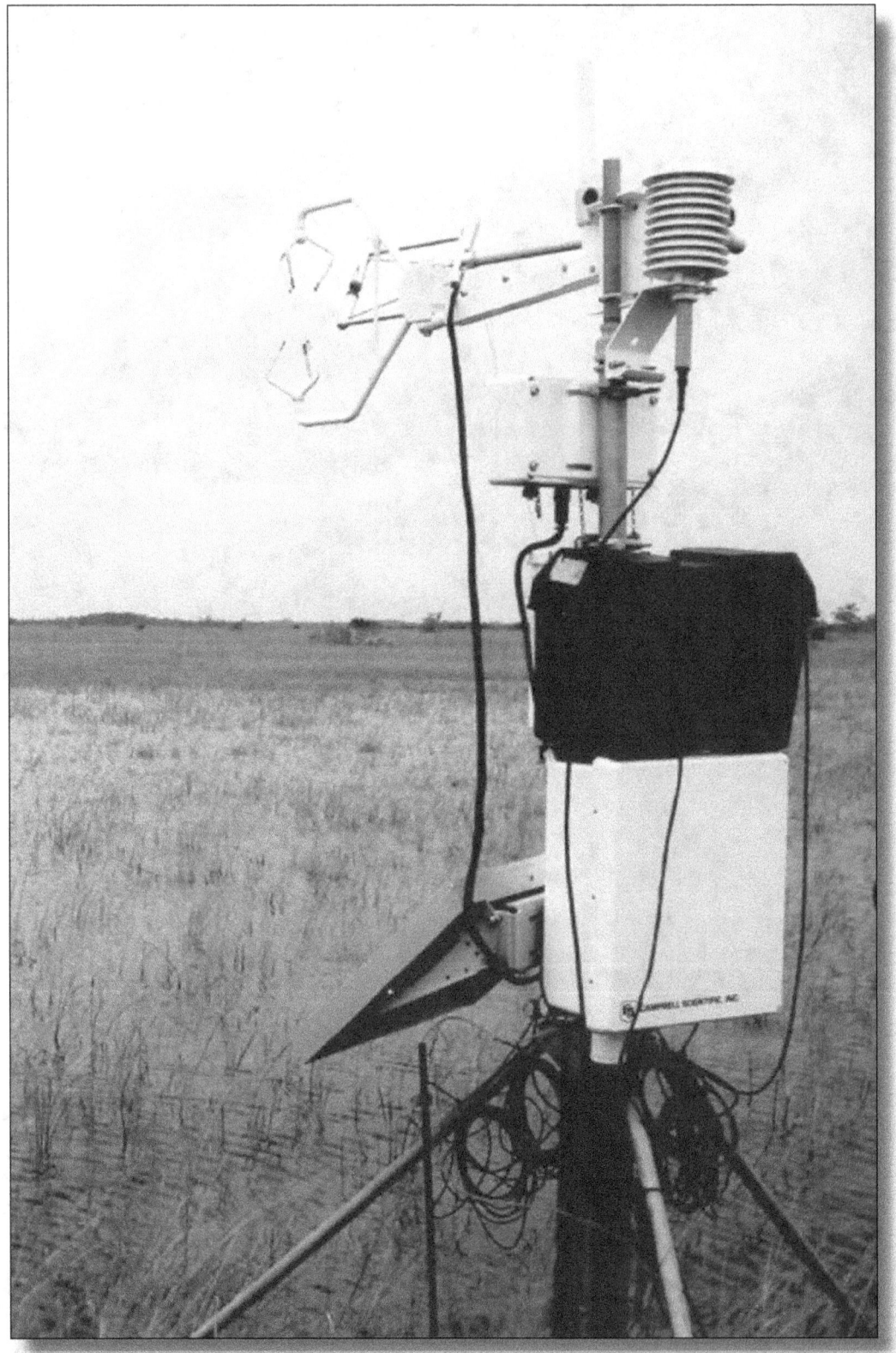

Figure 3.   Photographs of the evapotranspiration monitoring stations. Photograph by Patrick Lynch, South Florida Water Management District.—Continued

Table 1. Evapotranspiration site names, locations, tower heights, and vegetation and substrate description.

Site name (fig. 1)	Latitude	Longitude	Height of tower (meters)	Vegetation and substrate
Dwarf Cypress	25°45'45"	80°54'27"	16.5	Dwarf cypress and sawgrass (herbaceous vegetation)
Cypress Swamp	25°45'10"	81°06'01"	38	Tall cypress strand
Pine Upland	25°59'59"	80°55'29"	38	Pine upland and cypress domes
Wet Prairie	25°44'41"	80°56'24"	3.6	Wet prairie with short (about 1 meter) sawgrass (herbaceous vegetation
Marsh	26°11'57"	81°15'58"	3.6	Deep-water marsh with tall (about 1-2 meters) sawgrass (herbaceous vegetation

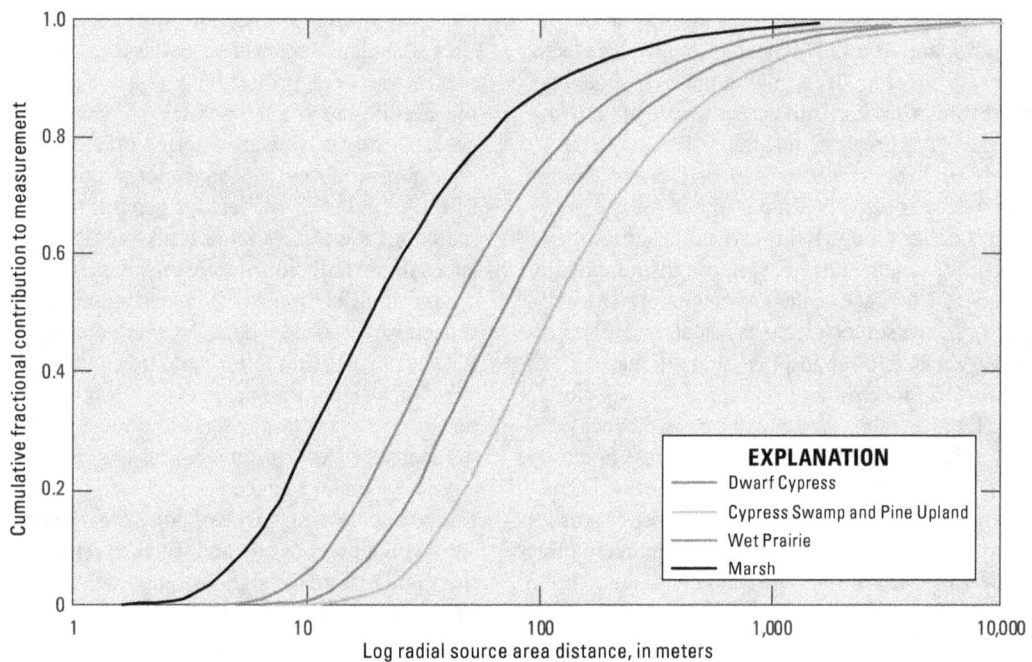

Figure 4. Radial extent of turbulent-flux source areas.

of the canopy height), and a "mildly unstable" atmosphere with an Obukhov length equal to -10 (fig. 4). At night, the atmosphere is likely stable, creating a substantially larger source area than 400 m. During the day, however, when ET is greatest, the atmosphere could be "mildly unstable" to "very unstable" with an Obukhov length equal to -10 and -1, respectively. Radial source areas would be substantially smaller (about 200 m) when the atmosphere is "very unstable." Thus, the eddy covariance instrumentation was mounted about 6 m above the 10-m canopy (fig. 4), so the source area was cypress under "mildly unstable" to "very unstable" atmospheric conditions.

A location west of Drill Pad 4 of Raccoon Point was chosen for the Pine Upland ET site (table 1 and fig. 1). At this site, 18- to 24-m pine trees interspersed with cypress domes are present for many kilometers surrounding Pad 4. According to Scheupp and others (1990), sensors placed about 12 m

above a 24-m pine canopy height create a radial source area extending about 800 m from the tower, assuming a roughness length of 2.4 m (about 10 percent of the canopy height), a displacement height of 15.6 m (about 65 percent of the canopy height), and a "mildly unstable" atmosphere with an Obukhov length equal to -10. Drill Pad 4 is a cleared area of about 24,000 m² within the Pine Upland source area. Impacts of this clearing on measured ET rates are expected to be minor, because the drill pad is about 1 percent of the expected source area with a "mildly unstable" atmosphere, and 5 percent of the expected source area with a "very unstable" atmosphere. At night, the atmosphere is likely stable, and the footprint extends much farther than 800 m. During the day, however, when ET is greatest, the atmosphere could be "mildly unstable" to "very unstable" with an Obukhov length equal to -10 and -1, respectively. Radial source areas would be substantially smaller (about 400 m) when the atmosphere is "very unstable." Thus, the eddy covariance instrumentation was mounted about 12 m above the mean canopy height on a 38-m tower (fig. 4), so the source area was pine under "mildly unstable" to "very unstable" atmospheric conditions.

East-Hensen Slough was selected for the Marsh ET site (table 1 and fig. 1). Reliable measurement of Marsh ET required construction of a 3.6-m tower for air sampling at distances about 2 m above the sawgrass canopy. A tall open-to-dense herbaceous plant community extends at least 250 m radially in every direction from the Marsh ET site. The tower was about 3.6 m. According to Scheupp and others (1990), sensors placed 2.6 m above 1 m canopy create a radial source area extending about 200 to 300 m from the tower, assuming a roughness length of 0.1 m (about 10 percent of the canopy height), a displacement height of 0.7 m (about 65 percent of the canopy height), and a "mildly unstable" atmosphere with an Obukhov length equal to -10. At night, the atmosphere is likely stable creating a substantially larger source area. During the day, when ET is greatest, the atmosphere could be "mildly unstable" to "very unstable" with an Obukhov length equal to -10 and -1, respectively. Radial source areas would be substantially smaller (about 100 m) when the atmosphere is "very unstable." Thus, the eddy covariance instrumentation was mounted about 2.6 m above the approximately 1-m-tall sawgrass (fig. 4), so the source area was Marsh under "mildly unstable" to "very unstable" atmospheric conditions.

An area west of Pinecrest on Loop Road was selected for the Wet Prairie ET site (table 1 and fig. 1). Reliable measurement of Wet Prairie ET required construction of a 3.6-m tower for air sampling at distances about 2 m above the low herbaceous canopy (fig. 4). A diverse, low ground cover of grasses, sedges, and forbs extends about 500 m radially in every direction from the Wet Prairie ET site. The tripod was about 2.5 m high. According to Scheupp and others (1990), sensors placed 1.5 m above a 1-m tall canopy create a radial source area extending about 200 m from the tripod, assuming a roughness length of 0.1 m (about 10 percent of the canopy height), a displacement height of 0.7 m (about 65 percent of the canopy height), and a "mildly unstable" atmosphere with an Obukhov length equal to -10. At night, the atmosphere is likely stable creating a much larger source area than 200 m. During the day, however, when ET is greatest, the atmosphere could be "mildly unstable" to "very unstable" with an Obukhov length equal to -10 and -1, respectively. Radial source areas would be substantially smaller (about 100 m) when the atmosphere is "very unstable." Thus, the eddy covariance instrumentation was mounted about 2.6 m above the roughly 1-m-tall sawgrass, so the source area was Marsh under "mildly unstable" to "very unstable" atmospheric conditions (fig. 4).

Instrumentation and Site Maintenance

Eddy covariance instrumentation comprises sonic anemometers and krypton hygrometers that measure latent- and sensible-heat fluxes, respectively (table 2). Hygrometer

voltage is proportional to attenuated radiation emitted from a hygrometer source tube to a hygrometer detector tube. Voltage fluctuations are related to fluctuations in vapor density by Beer-Lambert Law (Weeks and others, 1987). Sonic anemometers detect changes in the transit time of emitted sound waves to infer fluctuations in wind speed in three orthogonal directions, sonic air temperature, and sensible heat.

Meteorological instrumentation (fig. 5 and table 2) was installed to measure solar radiation, net radiation, soil-heat flux, vapor-density fluctuations, rainfall, soil-moisture content, air and soil temperatures, relative humidity, distance of water above and below land surface, and mean wind speed and direction and the maximum wind gusts measured during a 30-minute averaging period. Four component net radiometers were installed in April 2008 at the forested ET sites (Pine Upland, Cypress Swamp, and Dwarf Cypress) as part of a separate project with goals of quantifying albedos for satellite-based ET estimates. However, Kipp and Zonen net radiometers were used during quality-assurance quality-control procedures that force energy-budget closure and within Priestly-Taylor ET models for gap filling. Kipp and Zonen net radiometers were available at all five ET sites. Thus, differences in net radiation among the sites cannot be explained by use of different net radiometer brands and manufacturing procedures. Sap-flow data also were collected at the forested ET sites in an effort to estimate transpiration.

Sites visits were made every month to download data, perform a sensor inspection and other complete equipment maintenance. All instrumentation was visually inspected, leveled, cleaned, or replaced as necessary. Krypton hygrometer source and detector tube windows were cleaned when necessary with a cotton swab and water to remove dust obstructions and restore the signal strength. Net radiometers were releveled, if necessary. Desiccants were replaced to prevent moisture accumulation within instrumentation enclosures. Depth-to-water measurements were taken with a steel or electric water tape from the top of well casings to the water surface. Depth-to-water measurements were used to develop drift corrections for the pressure transducer readings. Digital photographs of the vegetation community were generally taken during monthly site inspections.

Priestley-Taylor Evapotranspiration Models

Dirty or water-obscured hygrometer lenses can result in turbulent flux data loss. Missing 30-minute and daily latent-heat fluxes were gap filled using Priestley-Taylor ET models prior to computing daily, monthly, and annual ET totals. Daily latent-heat fluxes were culled if greater than 50 percent of the 30-minute daytime latent-heat values were missing rather than measured. The culled daily latent-heat fluxes were subsequently gap filled with a Priestley-Taylor ET model constructed at daily time scales.

The Priestley-Taylor equation (Priestley and Taylor, 1972) estimates evaporation and assumes an extensive wet

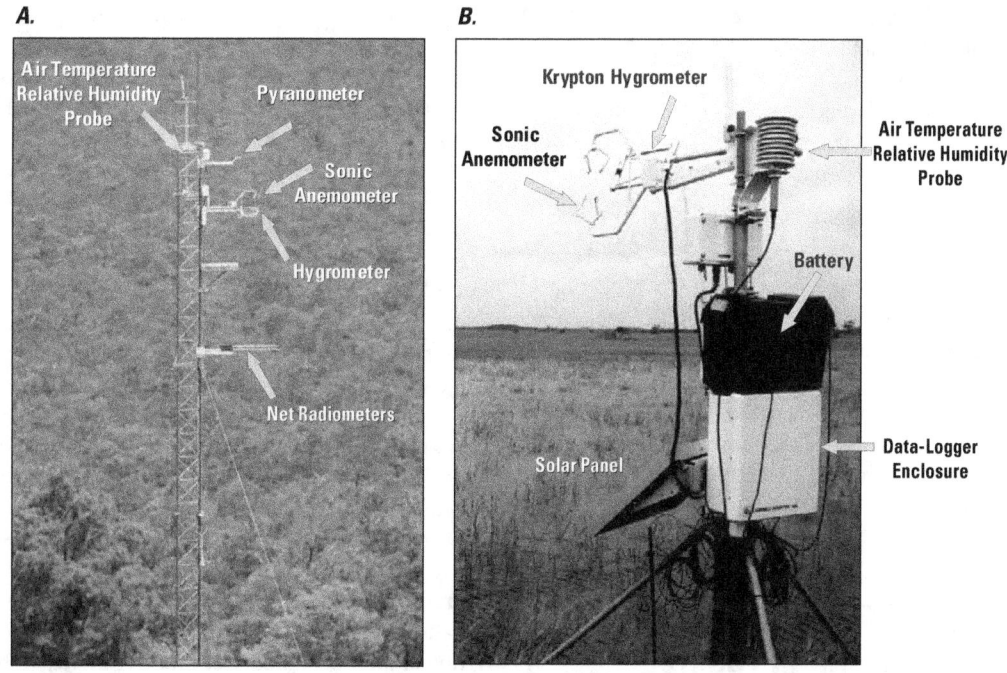

Figure 5. Photographs of (*A*) eddy covariance and (*B*) weather monitoring instrumentation.

Table 2. Study instrumentation.

Instrument	Measurement use	Distance installed above or below (-) land surface, in meters				
		Marsh	Wet Prairie	Dwarf Cypress	Cypress Swamp	Pine Upland
Sonic anemometer	Wind speed and direction	3.7	2.1	15.5	35.7	35.8
Krypton hygrometer	Vapor density fluctuations	3.7	2.1	15.5	35.7	35.8
Pressure transducer	Water distance	-.5	-.3	-.8	-.5	-.5
Air temperature/relative humidity probe	Air temperature/relative humidity	3.8	2.4	15.8	36.7	36.7
Net radiometer	Net radiation, R_n	3.4	2.4	13.2	33.9	33.7
Pyranometer	Incoming solar radiation	3.5	2.7	16.1	36.7	36.7
Rain gage	Rainfall	1.9	1.4	6.2	1.2	1.2
Wind sensor	Wind speed and direction	3.0	NA	11.4	30.6	NA
Soil heat-flux plates	Soil heat flux, G	-.2	-.2	-.2	-.2	-.2
Soil temperature probe	Soil temperature, S	-.1	-.1	-.1	-.1	-.1
Volumetric water content	Soil moisture	NA	-0.1	NA	-.1	-.1

surface under conditions of minimum advection. An alternate method, the Penman-Monteith equation (Monteith, 1965), incorporates additional processes that explain ET variability, such as atmospheric and plant stomata resistance to vapor transport. The additional processes, however, require additional data that are generally unavailable such as stomatal resistances. Stannard (1993) noted that a modified Priestley-Taylor approach for simulating ET was superior to the Penman-Monteith approach for a sparsely vegetated site in the semi-arid rangeland of Colorado. Sumner (1996) concluded the modified Priestley-Taylor approach performed better than Penman-Monteith at a site of herbaceous, successional vegetation in central Florida. Shoemaker and Sumner (2006) observed similar performance between modified Priestley-Taylor and Penman equations for simulating ET measured with Bowen ratios in the Everglades (German, 2000). Thus, a Priestley-Taylor equation was created for gap filling in this study. The Priestley-Taylor equation was formulated as:

$$\lambda E = \alpha \frac{\Delta}{\Delta + \gamma}(A_e)$$ (16)

where Δ is the slope of the saturated vapor pressure with respect to air temperature, in kilopascals per degrees Celsius; γ is the psychrometer constant equal to 0.07 kilopascal per degree Celsius, and α is a regression-defined coefficient that minimized residuals between measured and simulated 30-minute and daily latent-heat flux.

Quantifying Evapotranspiration in the Big Cypress National Preserve

Calibration methods and gap-filling results from the Priestley-Taylor model are presented and summarized. Selected annual and monthly means of the water- and energy-balance data also are presented as well as ET responses to hydrologic fluctuations and seasonal energy.

Application of Evapotranspiration Models

The regression-defined ET models were adequate for computing 30-minute and daily latent-heat fluxes (table 3). Results include (1) data availability for Priestley-Taylor model calibration, (2) model error statistics, and (3) the variability of the Priestly-Taylor α calibration coefficient defined by regression to minimize the sum-of-squared residuals between observed and predicted latent-heat fluxes. Also presented are graphical plots of model residuals in relation to observed values of latent-heat flux. These plots were helpful for identification of model bias over the range of observed values.

Data Availability for Model Calibration

Corrections and filtering of the eddy covariance data limited the number of observations available for Priestly-Taylor model calibration. Given a 30-minute time step, 17,520 measurements of latent-heat flux may be collected during a year; about half are measured at night when the friction velocity (u*) is usually less than 0.3. Therefore, one-half of the latent-heat data measurements are removed by the u* filter and are not available for calibration purposes. Measured latent heat also is lost to moisture or dust accumulating on the sampling lenses of the krypton hygrometer. Monthly cleaning of the lenses with a cotton swab and water restores signal strength.

Error Statistics and Model Efficiency

Error statistics for the Priestly-Taylor models (table 3) are adequate given the observed variability in latent-heat flux. For example, latent-heat flux varies from 0 to more than 300 W/m², whereas mean absolute residuals are about 37 and 10 W/m² for the 30-minute and daily ET models, respectively (table 3). Nash-Sutcliffe (Nash and Sutcliffe, 1970) model efficiency coefficients have been computed for each Priestly-Taylor ET model and can range from 0.0 to 1.0. An efficiency of 1.0 indicates the predicted latent-heat fluxes exactly represent observed latent-heat fluxes. Values progressively less than 1.0 indicate greater error in the Priestly-Taylor model calibration. Nash-Sutcliffe coefficients for the Priestly-Taylor models are generally greater than about 0.7 (table 3), indicating that more than 70 percent of the observed variability in latent-heat flux is adequately represented by the Priestly-Taylor ET models.

The Marsh and Wet Prairie ET models were most problematic. Specifically, about 40 to 150 observed daily values of latent-heat flux were available for model calibration at the Marsh and Wet Prairie ET sites (table 3) in years 1 to 3. The Marsh ET site daily mean absolute residual is 21 W/m² for year 3. The Nash-Sutcliffe coefficient was 0.13 and 0.46 for the daily ET model at the Marsh site in years 2 and 3, respectively. Krypton hygrometers failed more frequently for unknown reasons at the Marsh and Wet Prairie sites. Hygrometer failure is difficult to identify because slow signal loss translates into minor downward drift of latent heat over a period of several months. Hygrometer failures likely explain the lack of data and large mean absolute residuals at the Wet Prairie and Marsh ET sites during the second and third measurement years.

Factors Limiting Evapotranspiration Rates

Priestley and Taylor (1972) estimated a value of 1.26 for α over a free-water surface or a dense, well-watered canopy. Conversely, values of 1.26 overestimate ET by as much as 100 percent in the Everglades (Shoemaker and Sumner, 2006, table 4). At BCNP ET field sites, regression-defined estimates of α range from 0.55 to 1.07 (table 3), and are similar in magnitude to α in the Everglades (Shoemaker and Sumner, 2006, appendix 1). When water is readily available at the sites, ET generally was not occurring at potential rates. Apparently, factors other than water availability also are limiting the ET rates, for example, atmospheric or stomatal resistance to vapor transport. There also is some evidence of vegetational limiting of ET at relatively large vapor-pressure deficits. Specifically, studies of transpiration in cypress trees in southern Florida suggest that even though water is readily available, cypress transpiration is limited through closure of stomata during photosynthesis at relatively large vapor pressure deficits (Brian Bovard, Florida Gulf Coast University, written commun., 2005). Further data are being collected to assess and confirm the magnitude of this possible ET limiting factor.

Model Bias

As noted previously, daily latent-heat flux residuals (fig. 6) help identify Priestly-Taylor model bias. At the Dwarf Cypress site (fig. 6), for example, the daily Priestly-Taylor models overestimate observed latent-heat flux between 50 and 100 W/m². A bias also occurred when observed latent-heat flux is greater than 100 W/m². Specifically, the Priestly-Taylor model underestimates observed latent-heat flux that was greater than 100 W/m². At the Wet Prairie site in year 2 (fig. 6), the Priestly-Taylor model underestimates observed latent-heat flux greater than 100 W/m². At the Pine Upland site (fig. 6), the Priestly-Taylor models appear biased when observed latent-heat flux is greater than 50 W/m². Specifically, the residuals approached zero as observed latent-heat flux increase from 50 and 100 W/m². Residuals become increasingly positive as observed latent-heat flux exceed 100 W/m², indicating the Priestly-Taylor model under predicts the observed latent-heat fluxes that exceed 100 W/m². At the Cypress Swamp site (fig. 6), the Priestly-Taylor models generally overestimate observed latent-heat fluxes less than 100 W/m², and underestimate observed latent-heat fluxes

greater than about 150 W/m². At the Marsh site (fig. 6), residuals appear randomly distributed when observed latent-heat flux is less than about 100 W/m². The Priestly-Taylor model tends to underestimate observed latent-heat flux greater than 100 W/m² at the Marsh site, as indicated by the positive residuals. The Priestly-Taylor model biases may be removed through calibration of more complex ET equations as described by Shoemaker and Sumner (2006), which was beyond the scope of work for this study.

Annual Water- and Energy-Budget Calculations

Annual water- and energy-balance statistics were determined at the five ET sites (table 4). Summarized data include rainfall, soil volumetric water content, net radiation, hydroperiod, air temperature, water distance from land surface, latent-heat flux, sensible-heat flux, the Bowen ratio, wind speed, maximum wind gust, ET, available water computed as the difference between rainfall and ET, evaporative fraction, and energy-budget closure. Bowen ratios were computed as the ratio of mean annual sensible-heat flux to latent-heat flux.

Table 3. Error statistics for the 30-minute and daily Priestly-Taylor models at the evapotranspiration sites.

[α, regression-defined Priestly-Taylor alpha coefficient; W/m², watts per square meter]

Year	Time period	30-minute models				Daily models			
		No. of observations	α	Mean absolute error (W/m²)	Nash-Sutcliffe Model efficiency coefficient	No. of observations	α	Mean absolute error (W/m²)	Nash-Sutcliffe model efficiency coefficient
					Dwarf Cypress				
1	10/10/07 – 10/09/08	5,336	0.76	32	0.87	294	0.79	7	0.90
2	10/10/08 – 10/09/09	3,697	.75	37	.79	183	.77	9	.60
3	10/10/09 – 10/09/10	2,792	.78	38	.82	137	.72	10	.79
					Wet Prairie				
1	10/10/07 – 10/09/08	2,715	0.79	35	0.84	145	0.80	9	0.83
2	10/10/08 – 10/09/09	812	1.07	28	.92	45	1.03	7	.82
3	10/10/09 – 10/09/10	2,184	.81	29	.90	113	.87	5	.97
					Pine Upland				
1	10/10/07 – 10/09/08	4,052	0.71	37	0.80	249	0.72	6	0.87
2	10/10/08 – 10/09/09	3,992	.65	43	.73	223	.68	14	.59
3	10/10/09 – 10/09/10	4,350	.72	36	.84	241	.78	9	.82
					Cypress Swamp				
1	10/10/07 – 10/09/08	4,655	0.84	43	0.84	324	0.59	9	0.92
2	10/10/08 – 10/09/09	4,593	.75	51	.73	239	.79	10	.83
3	10/10/09 – 10/09/10	4,468	.77	45	.79	223	.81	12	.83
					Marsh				
1	10/10/07 – 10/09/08	2,883	0.68	32	0.84	119	0.67	10	0.76
2	10/10/08 – 10/09/09	2,079	.55	33	.74	87	.60	13	.13
3	10/10/09 – 10/09/10	2,351	.90	32	.92	96	.73	21	.46

Table 4.   Annual water- and energy-balance calculations at the evapotranspiration sites, October 10, 2007 to October 9, 2010.

[Year 1: October 10, 2007 to October 9, 2008; year 2: October 10, 2008 to October 9, 2009; year 3: October 10, 2009 to October 9, 2010. BCNP, Big Cypress National Preserve. NA, Data not available. Daily NEXRAD data (rainfall, in millimeters) for the EDEN, Everglades Depth Estimation Network, gage locations were compiled from 15-minute NEXRAD data provided by the South Florida Water Management District. The end-of-month data have been verified and are documented in appendix 2-1 (p. 57) of the 2008 South Florida Environmental Report Volume 2 available at *http://my.sfwmd.gov/SFER*. The Dwarf Cypress and Wet Prairie rainfall data were compiled from the BCNP LOOP1 monitoring station. The Cypress Swamp, Pine Upland, and Marsh rainfall data were compiled from the BCNP BCA11, BCA5, and BCA2 monitoring stations, respectively]

Balance component	Dwarf Cypress			Wet Prairie			Cypress Swamp			Pine Upland			Marsh		
	Year 1	Year 2	Year 3	Year 1	Year 2	Year 3	Year 1	Year 2	Year 3	Year 1	Year 2	Year 3	Year 1	Year 2	Year 3
Total rainfall, in millimeters	1,448	1,219	1,270	1,448	1,219	1,270	1,600	1,194	1,372	1,524	1,118	1,219	1,651	1,346	1,321
Mean soil volumetric water content, ratio of saturated to unsaturated porosity	NA	NA	NA	.63	.65	.66	.58	.50	.58	.29	.25	.36	NA	NA	NA
Mean net radiation, in watts per square meter	113	119	115	115	124	117	128	133	128	114	122	115	111	125	118
Hydroperiod, in number of flooded days per year	345	325	346	94	108	108	343	251	346	120	80	29	65	179	352
Mean air temperature, in degrees Celsius	24.2	21.7	23.4	23.7	23.7	23.0	24.0	23.2	23.6	24.2	23.3	23.5	22.8	22.3	23.5
Water distance from above and below (-) land surface	.14	.15	.13	-.08	-.07	-.06	.12	.06	.13	-.09	-.22	-.10	-.16	-.01	.17
Mean latent-heat flux, in watts per square meter	76	81	68	79	79	86	87	82	74	65	65	73	63	65	83
Mean sensible heat flux, in watts per square meter	56	60	60	51	61	51	54	67	57	60	68	63	67	60	66
Mean Bowen ratio, unitless	.75	.74	.89	.64	.77	.59	.62	.82	.77	.93	1.05	.86	1.06	.93	.79
Mean wind speed, in meters per second	2.6	2.7	2.5	NA	NA	NA	2.8	2.7	2.7	NA	NA	NA	1.7	1.5	1.6
Maximum wind gust, in kilometers per hour	62	70	97	NA	NA	NA	78	72	86	NA	NA	NA	102	73	57
Total evapotranspiration, in millimeters	976	1,075	959	1,017	1,019	1,106	1,179	1,115	1,025	876	909	996	816	840	1,068
Available water (rainfall – evapotranspiration), in millimeters	471	144	311	430	201	164	421	79	347	648	209	223	835	506	253
Evaporative fraction, unitless	.67	.68	.59	.68	.64	.73	.68	.61	.57	.56	.54	.63	.57	.52	.70
Energy-budget closure, in percent	73	69	74	98	81	90	83	81	85	82	80	80	83	81	75

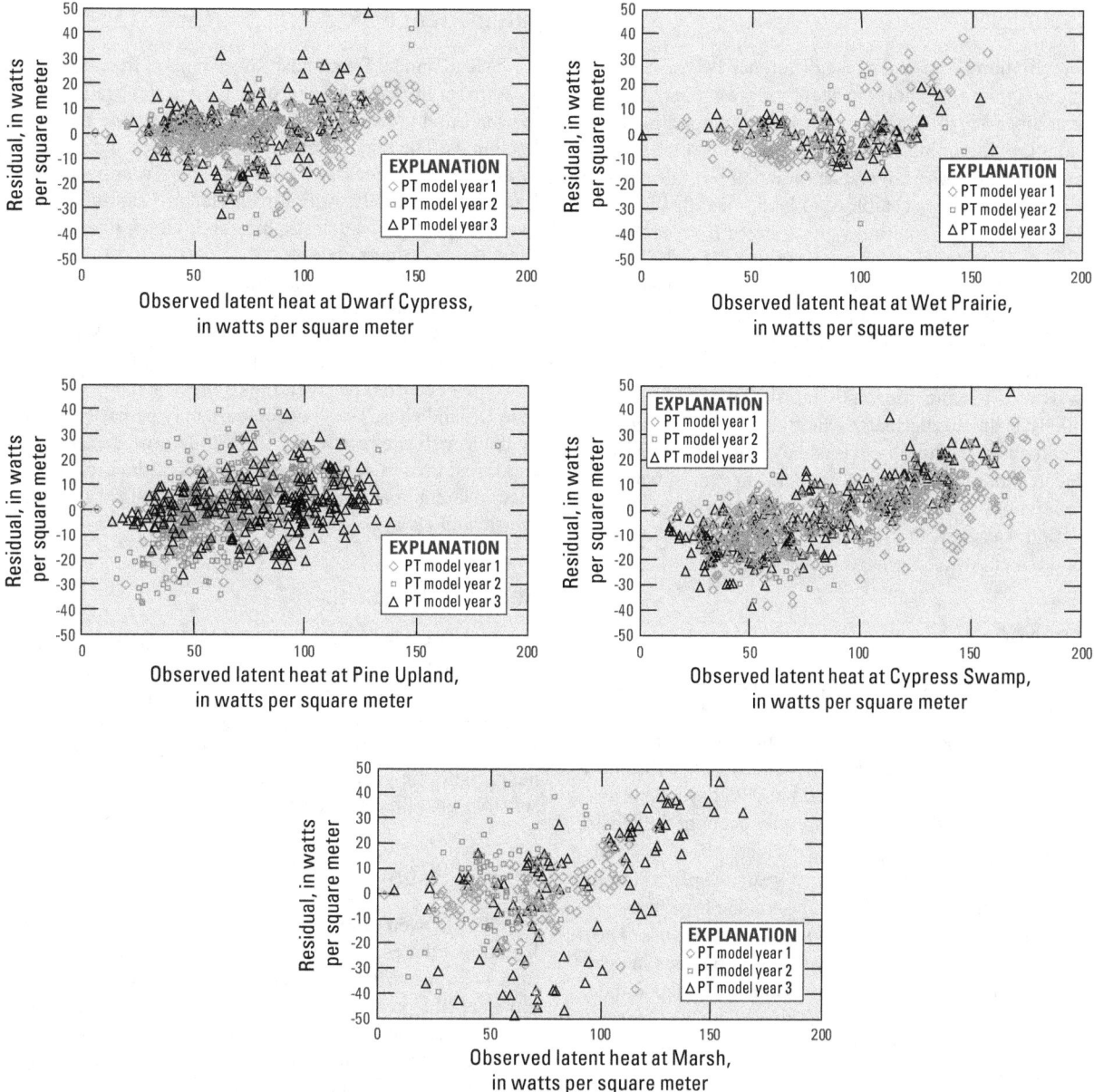

Figure 6. Observed latent heat in relation to residuals for the daily Priestly-Taylor (PT) evapotranspiration models.

Evaporative fractions were computed as the ratio of latent-heat flux to net radiation. The energy-budget closure statistic was computed as the percent of available energy measured by the eddy covariance instrumentation. For example, 80-percent energy-budget closure indicates the sum of the eddy covariance latent- and sensible-heat fluxes was 80 percent of the measured available energy.

Rainfall

The distribution, frequency, and intensity of rainfall vary substantially in southern Florida. The humid wet season extends from May to October, with rainfall occurring as

a result of convective thunderstorms and tropical cyclones. The mild dry season extends from November to April with relatively little rainfall.

Annual rainfall totals range from 1,118 to 1,651 mm/yr at the five sites and were least at the Pine Upland site in year 2 and greatest at the Marsh site in year 1 (table 4). Year 1 was a relatively wet year with rainfall totals ranging from 1,448 mm (Dwarf Cypress and Wet Prairie sites) to 1,651 (Marsh site). Year 2 was a relatively dry year with rainfall totals ranging from 1,118 mm (Pine Upland site) to 1,346 mm (Marsh site). Year 3 was a relative moderate rainfall period, with totals ranging from 1,270 mm (Dwarf Cypress and Wet Prairie sites) to 1,372 mm (Cypress Swamp site).

Net Radiation

Net radiation is defined as the difference between the incoming and outgoing shortwave and long-wave radiation. Terrestrial albedo is the fraction of reflected (outgoing) short-wave radiation. Incoming shortwave radiation is a function of solar radiation and cloud cover. Net radiation provides most of the available energy for sensible- and latent-heat fluxes.

Mean annual net radiation values ranged from 111 to 133 W/m² at the ET sites and were least at the Marsh site in year 1 and greatest at the Cypress Swamp site in year 2 (table 4). Mean annual net radiation was least at all of the ET sites (averaging about 116 W/m²) in year 1, and least at the Dwarf Cypress site (averaging about 115 W/m²) over 3 years. The Dwarf Cypress site apparently has the largest albedo, which likely is due to the sparse canopy and a highly reflective, calcareous, periphyton-covered land surface. Furthermore, mean annual net radiation was least in year 1, which was a relatively wet year with more cloud cover. In contrast, net radiation was greatest in year 2, which likely was due to less cloud cover during a relatively dry year.

Hydroperiod

A hydroperiod is defined as the number of days per year that the mean daily water level is above land surface at a site. The hydroperiod ranged from 29 to 352 days at the ET sites (table 4). The Wet Prairie and Pine Upland sites had the shortest hydroperiods, averaging about 103 and 76 days, respectively. The longest hydroperiods generally occur at the Dwarf Cypress and Cypress Swamp sites, averaging about 339 and 313 days per year, respectively, over 3 years. Cypress strands and domes generally occur within topographic lows and sloughs, which partly explains the longer hydroperiods. The hydroperiod at the Marsh site increased from 65 days in year 1 to 352 days in year 3. Drought conditions prevailed at the Marsh site in 2007 prior to ET site construction. The Marsh site flooded in August 2008 as rainfall began to alleviate the drought condition.

Air Temperature

Mean annual air temperature varied within about 2 °C over the 3-year study period. (table 4). Mean annual air temperatures were 24.0 and 23.6 °C in years 1 and 3, respectively, at the Cypress Swamp site. Additionally, mean annual air temperatures were about 22.8 and 23.5 °C in years 1 and 3, respectively, at the Marsh site. Lengthy cold fronts in years 2 and 3 may explain the apparent decrease in mean annual air temperature at all five ET sites over 3 years. For example, Florida experienced an extreme cold front on or around January 1, 2010, when air temperatures of about 5 °C were measured at the ET sites every night for about 1 week. This extreme cold front had consequences for the ecology of the BCNP and subtropical southern Florida, including the death of many exotic and native species.

Latent-Heat Flux

Mean annual latent-heat fluxes ranged from 63 to 87 W/m² at the ET sites and were least at the Marsh site in year 1 and greatest at the Cypress Swamp site in year 1 (table 4). The maximum mean annual latent-heat fluxes of 87 and 82 W/m² occurred at the Cypress Swamp site in years 1 and 2, respectively. Additionally, latent-heat flux was relatively large at the Marsh site in year 3, which likely was due to a lengthy hydroperiod (table 4).

Spatial differences in latent-heat flux were considerable on an annual basis. In year 1, the greatest spatial difference (about 24 W/m²) occurred between the Cypress Swamp and Marsh sites. In year 2, the greatest spatial difference (about 17 W/m²) occurred between the Cypress Swamp and Marsh/Pine Upland sites. These differences may be partly explained by the Marsh site recovering from 2007 drought conditions and forest fires in years 1 and 2. In year 3, the greatest difference in latent heat (about 18 W/m²) occurred between the Wet Prairie and Dwarf Cypress sites.

Sensible-Heat Flux

Sensible-heat fluxes ranged from 51 to 68 W/m² at the ET sites and were greatest at the Pine Upland site in year 2 and least at the Wet Prairie site in years 1 and 3 (table 4). Sensible-heat flux was mostly greater in year 2 than in years 1 and 3 at the ET sites, which perhaps is due to greater annual net radiation in year 2. The Marsh site was an exception where sensible-heat flux was least in year 2.

Bowen Ratio

The Bowen ratio (Bowen, 1926) is the ratio of sensible- to latent-heat fluxes. A relatively large Bowen ratio (i.e., greater than 1) indicates an ecosystem converts available energy into more sensible heat than latent heat. A relatively small Bowen ratio (i.e., less than 1) indicates an ecosystem converts available energy into more latent heat than sensible heat. Bowen ratios at the Pine Upland and Marsh sites were generally greater than those at the Dwarf Cypress, Wet Prairie, and Cypress Swamp sites (table 4). The greater Bowen ratios at the Pine Upland site may be due to less water available for evaporation. This evidence is supported by the decreasing Bowen ratio at the Marsh site with increasing hydroperiod (table 4).

Evapotranspiration

Similar to latent-heat flux, spatial differences in annual ET were considerable. Mean annual latent-heat fluxes were converted to ET using the density and latent heat of vaporization of water as well as the number of seconds in a year. Annual ET values ranged from about 800 to 1,200 mm/yr at the ET sites and were greatest at the Cypress Swamp site in year 1 and lowest at the Marsh site in year 1. The differences in ET between the Cypress Swamp and Marsh sites were 363 mm in year 1 and 275 mm in year 2.

The relatively small ET rates (816 mm in year 1 and 840 mm in year 2) at the Marsh site (table 4) are likely explained by the site recovering from an extensive forest fire coupled with drought conditions. Land surface was exposed at the Marsh site for most of year 1 and about half of year 2 due to a severe drought that persisted in southern Florida for many months. Exposed land surface is unusual at the Marsh site; the water distance above land surface is generally 0.5 to 1.0 m. Dry conditions facilitated the spread of a forest fire that burned about 486 ha (D.G. Doumlele, Big Cypress National Preserve, oral commun., 2007). Access to the Marsh site was prohibited for about 1 month, until fire fighters declared the area safe for work and recreation.

Evaporative Fraction

Evaporative fraction is defined as the ratio of latent-heat flux to net radiation. A relatively large evaporative fraction indicates net radiation creates mostly water vapor in the form of latent heat flux. A relatively small evaporative fraction indicates latent heat flux is a smaller portion of the surface energy budget. Mean annual evaporative fractions ranged from 0.52 to 0.73 and were largest at the Wet Prairie site in year 1 and smallest at the Marsh site in year 2 (table 4). Mean annual evaporative fractions were similar from year to year at the ET sites. For example, the evaporative fractions equaled 0.67 in year 1, 0.68 in year 2, and 0.59 in year 3 at the Dwarf Cypress site. Given this result, annual ET may be approximated reasonably well as the product of the net radiation and the average of the evaporative fractions reported herein (table 4), if surface-energy and water-availability conditions are similar to those reported herein.

Available Water

Available water is defined as the difference between annual rainfall and ET. This difference quantifies the water available for runoff to coastal estuaries and net recharge to the water table. Annual available water values ranged from 79 to 835 mm at the ET sites and were greatest at the Marsh site in year 1 and least at the Cypress Swamp site in year 2. The positive available water values indicate surplus rainfall was always available for aquifer recharge and sheet flow toward the coast on an annual basis.

Energy-Budget Closure

Energy-budget closure is the percent of available energy measured by the sum of the eddy covariance latent- and sensible-heat fluxes. Previous investigators (Lee and Black, 1993; Bidlake and others, 1996; Sumner, 1996; and German, 2000) have noted the sum of the measured latent- and sensible-heat fluxes is generally less than the measured available energy. Foken (2008) determined low-frequency (large-scale) eddies unmeasured by the high-frequency eddy covariance systems explain this discrepancy. Thirty-minute latent- and sensible-heat fluxes were summed over a year for comparison with 30-minute summations of available energy. Energy-budget closure ranged from 74 to 98 percent, with the highest percent occurring at the Wet Prairie site in year 1 and the lowest percent occurring at the Dwarf Cypress site in year 1 (table 4).

Monthly Water- and Energy-Budget Calculations

Monthly water- and energy-balance calculations were determined at the five ET sites from October 2007 to September 2010. Summarized data include rainfall, ET, available water, soil volumetric water content, net radiation, and air temperature. Available water was computed as the difference between monthly rainfall and ET. Surface-energy fluxes also were averaged, including net radiation, latent- and sensible-heat fluxes. Bowen ratios were computed as the ratio of mean monthly sensible- to latent-heat flux. Evaporative fractions were computed as the ratio of mean monthly latent-heat flux to net radiation. All monthly statistics are presented in appendix 1 at the end of this report.

Rainfall

Monthly rainfall ranged from 4 to about 500 mm at the ET sites from October 2007 to September 2010 (fig. 7 and appendix 1). The greatest amount of rainfall occurred at the Pine Upland site (487 mm) and Marsh site (474 mm) in August 2008. All five ET sites experienced months with very little rainfall; specifically, November through January, which typically are the driest months of the year. The timing of the wet season varied (fig. 7), occurring from June 2008 to September/October 2008 in year 1 and from May 2009 to August/September 2009 in year 2. An unusually wet winter occurred at the ET sites starting in October 2009, with rainfall amounts ranging from 30 mm at the Pine Upland site in October 2009 to 111 mm at the Cypress Swamp site in January 2010.

Evapotranspiration

Monthly ET ranged from 37 to 143 mm at the ET sites over the 3-year period of record (fig. 8 and appendix 1). In years 1 and 2, monthly ET was greatest at the Cypress Swamp site and least at the Marsh site. In year 3, monthly ET was comparable at all five sites, which is likely due to the increased hydroperiod at the Marsh site during this time frame.

Seasonality was apparent in monthly ET, with rates generally greatest from March to October when solar radiation was relatively large, and least from November to February when solar radiation was relatively small. Monthly ET was greatest in May at the Cypress Swamp site, reaching rates as large as 150 mm. The large ET rates coincide with the annual leaf-in growth stage of the tall cypress before summer. Cypress trees reached full leaf-capacity before summer, while losing essentially all leaves in the early winter toward the end of hurricane season.

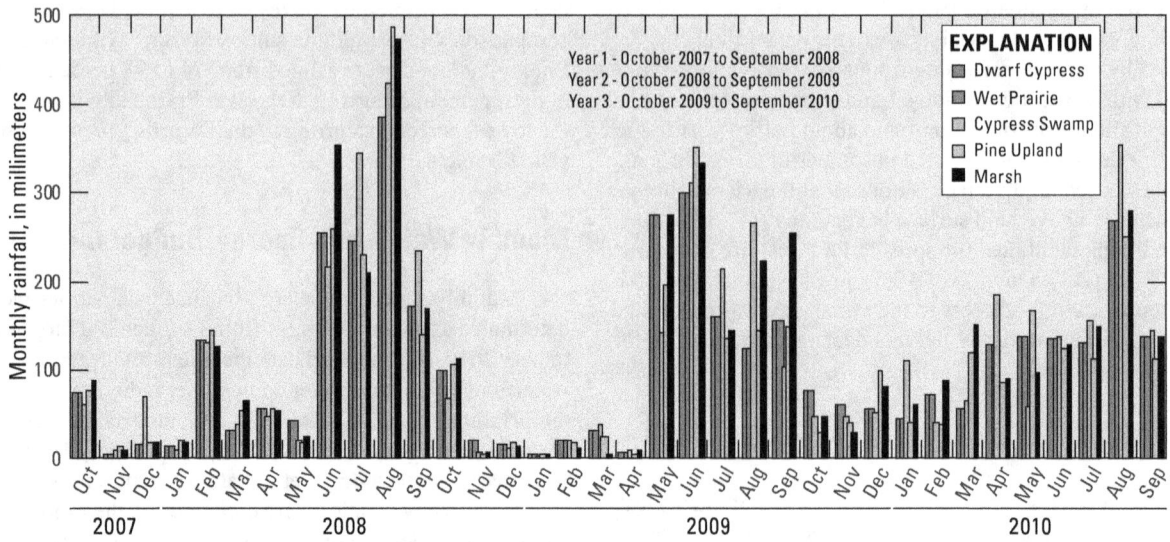

Figure 7. Monthly rainfall at the evapotranspiration sites, October 2007 to September 2010.

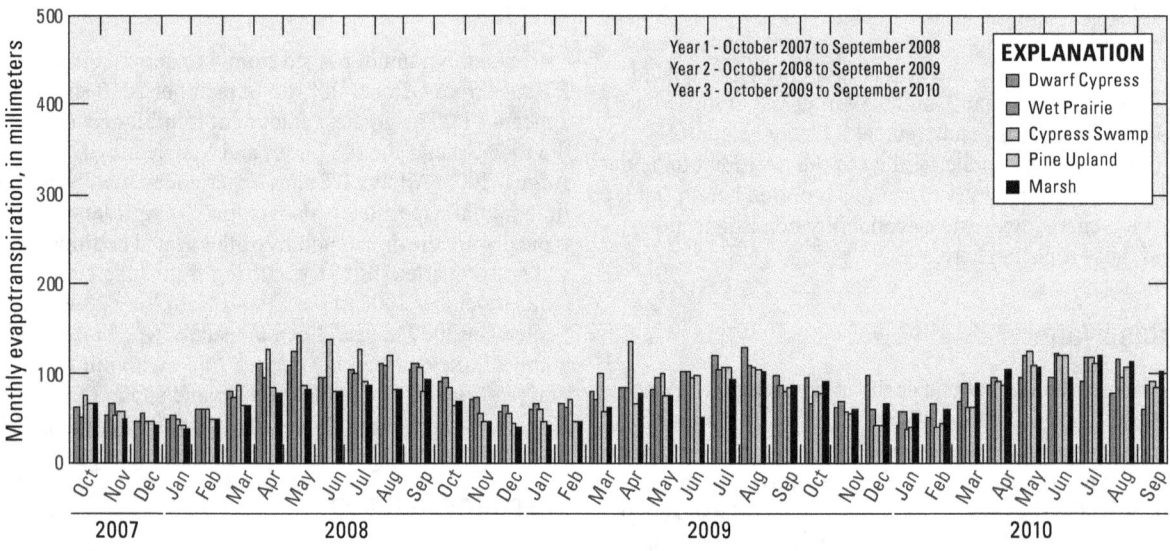

Figure 8. Monthly evapotranspiration (ET) at the ET sites, October 2007 to September 2010.

Available Water

Available water is computed as the difference between monthly rainfall and ET (fig. 9 and appendix 1). Monthly available water values ranged from -126 to 406 mm at the five sites between October 2007 to September 2010. The Pine Upland and Marsh site received the greatest amount of available water (more than 400 mm) in August 2008. The least amounts of available water were measured at the Cypress Swamp site in May 2008 (-121 mm) and

May 2009 (-126 mm). Negative available water generally occurred at all the sites in the relatively dry winter (October to May). Positive available water generally occurred at all the sites in the hot and humid summer (June to September) and appeared to be concentrated within 3 to 4 months (June to September) over a year. Available water was generally negative for the remainder of the year. Variations in available water were mostly explained by rainfall variability (fig. 7). The unusually wet winter starting in October 2009 is apparent in the available-water timeseries.

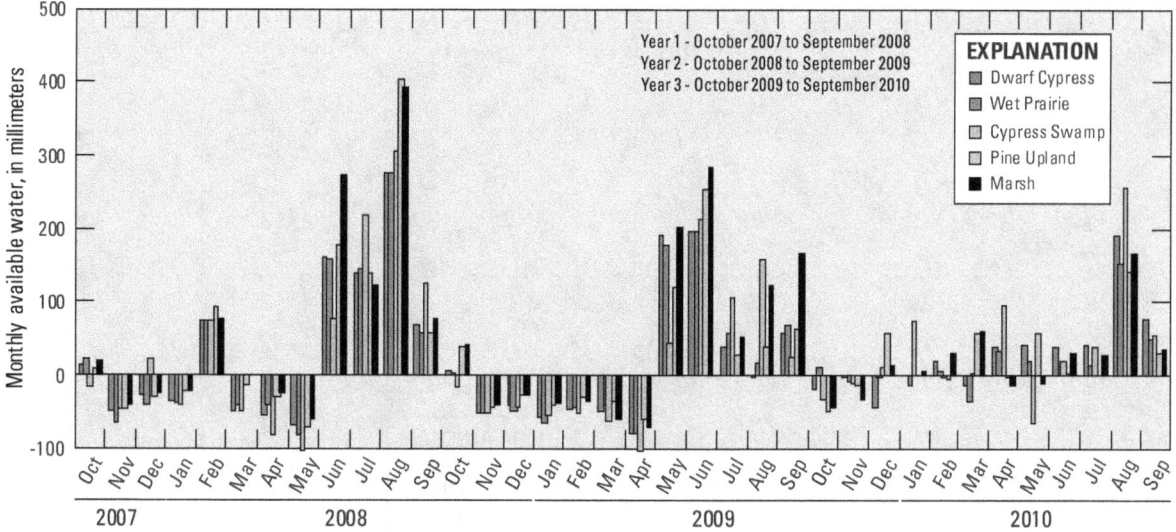

Figure 9. Monthly available water at the evapotranspiration sites, October 2007 to September 2010.

Soil Volumetric Water Content

Soil volumetric water content was measured only at the Wet Prairie, Cypress Swamp, and Pine Upland sites (see appendix 1). A constant value of 0.50 was assumed when computing changes in soil heat storage at the Marsh and Dwarf Cypress sites. The soil water content was greatest (0.46 to 0.72) in the organic muddy soils at the Wet Prairie site and least (0.04 to 0.37) in the sandy soils at the Pine Upland site. Soil water content decreased considerably at all three sites from about January to June 2009 due to limited rainfall.

Net Radiation

Monthly net radiation ranged from 58 to 195 W/m² at the ET sites from April 2007 to September 2010 (see appendix 1). Monthly net radiation was generally greatest at the Cypress Swamp site, averaging about 130 W/m² over the 3-year period of record. Monthly net radiation averaged about 120 W/m² each at the Dwarf Cypress, Wet Prairie, Pine Upland, and Marsh sites. Seasonality was apparent in monthly net radiation, which was greatest from March to October when incoming solar radiation was relatively large, and least from November to February when incoming solar radiation was relatively small.

Air Temperature

Mean monthly air temperature ranged from about 15 to 29 °C (fig. 10 and appendix 1). As expected, air temperatures were highest in the summer months (with values greater than 25 °C) and lowest in the winter months (with values less than 20 °C). A noteworthy trend was apparent in winter air temperatures over the 3-year period of record—the winters appeared to be progressively colder, which could have been due to climatic cycles, such as El Niño and La Niña.

Water and Energy Budget Calculations

Diurnal water-level variability in response to ET was observed at three ET sites, and the results are discussed herein. Additionally, the relative magnitude of each term comprising available energy is described, including net radiation, soil-heat flux, and changes in heat energy stored in the soil, surface water, and air column beneath the eddy covariance sensors. A seasonality index for ET is introduced, discussed, and compared at all five sites. Spatial ET differences between sites and temporal ET differences are examined as well as the ET response to water availability and the surface-energy budget during cold fronts.

Diurnal Water-Level Variability

Diurnal water-level variability in response to ET was less at flooded sites than at dry sites. The magnitude of these water-level declines is similar to the magnitude of shallow groundwater level declines due to transpiration within tree islands observed by Wetzel and others (2011) in the Everglades. Diurnal water-level variability and ET can be used to estimate specific yield to serve as a check on the accuracy of the eddy covariance method. Specific yield relates volumetric changes in fluid volume (per unit volume) to temporal changes in water levels. Sumner (2007) used specific yield to relate net depth-equivalent water fluxes to water-level changes as:

$$\Delta h = \frac{P - ET(Q_{in} - Q_{out})}{S_y}$$
(17)

where Δh is change in water level, in millimeters; P is precipitation, in millimeters; Q_{in} and Q_{out} are surface-water and groundwater inflow and outflow, in cubic meters per second,

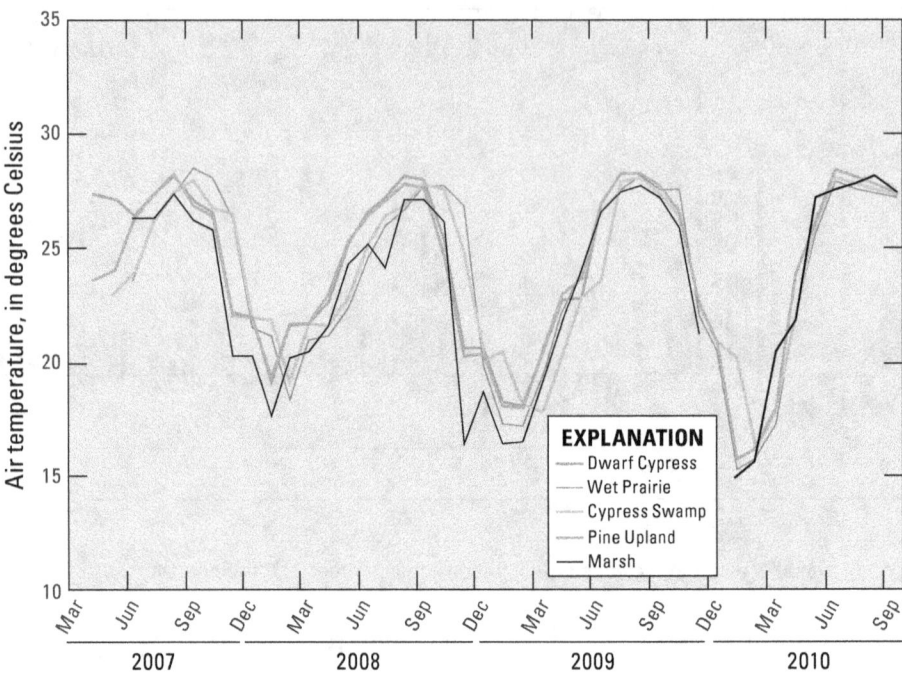

Figure 10.   Mean monthly air temperature at the evapotranspiration sites, March 2007 to September 2010.

respectively; and S_y is specific yield. Equation 17 can be simplified to estimate specific yield by isolating time periods when ET predominately explains water-level changes as:

$$S_y = \frac{ET}{\Delta h} \qquad (18)$$

Water levels repeatedly dropped about 15 mm on average in response to ET at the Pine Upland site from April 27 to May 4, 2008 (fig. 11C). ET was about 3 mm on each of these days, resulting in an estimated specific yield of 0.2. Note this estimate is similar to specific yields estimated by Merritt (1996) for shallow sandy limestones in the Biscayne aquifer, which ranged from 0.16 to 0.33. The geologic framework of the Pine Upland site also is characterized by sandy limestones.

Available Energy

Energy-budget terms were averaged into a single value every 30 minutes over the 3-year period of record. Results indicate the relative magnitudes of terms impacting available energy for latent and sensible heat (fig. 12). Available energy was generally less than net radiation until early-to-late afternoon, as a portion of net radiation was used to heat the soil, surface water, and air column. Available energy was generally greater than net radiation during the late afternoon and night as the energy absorbed to heat the soil, surface water, and air column was released back into the atmosphere, and thus, made available for latent and sensible heat. These results suggest relatively mild amounts of ET can occur well into the night as the landscape radiates energy away from land surface.

Changes in heat energy stored in the soil were questionable (fig. 12), likely due to measurement errors. Precision of soil-temperature probes may be about ±0.4 percent of the reading (Omega Engineering Inc., oral commun., 2005). The mean soil temperature measured at the Cypress Swamp site was about 25 °C, which translates into a possible error of about ±0.1 °C. An error of ±0.2 °C is possible for soil-temperature changes because these changes are the difference between two consecutive readings. An energy-flux error of about 30 W/m² results from an error of ±0.2 °C in soil-temperature change over 30 minutes for a 152-mm (6-in.) soil column. Changes in heat energy in the surface water were less noisy because water-temperature changes were computed as a convolution of air-temperature changes with a regression-defined transfer function.

At the Dwarf Cypress, Wet Prairie and Marsh sites, changes in heat energy in the surface water (W) and the soil heat-flux (G) were generally the largest energy fluxes with exception to net radiation, available energy, and latent- and sensible-heat fluxes (fig. 12). At the Cypress Swamp site, changes in heat energy in the surface water (W) and the air column (ΔA) were generally the largest energy fluxes except for net radiation, available energy, and latent- and sensible-heat fluxes (fig. 12). Humidity was routinely high within the dense Cypress Swamp canopy, which may explain the relatively large estimated energy-storage change in the air column (ΔA). At the Pine Upland site, change in heat energy in the surface water (W) was generally the largest energy flux except for net radiation, available energy, and latent- and sensible-heat fluxes. These flux relations are consistent with German (2000) and Shoemaker and Sumner (2005), who

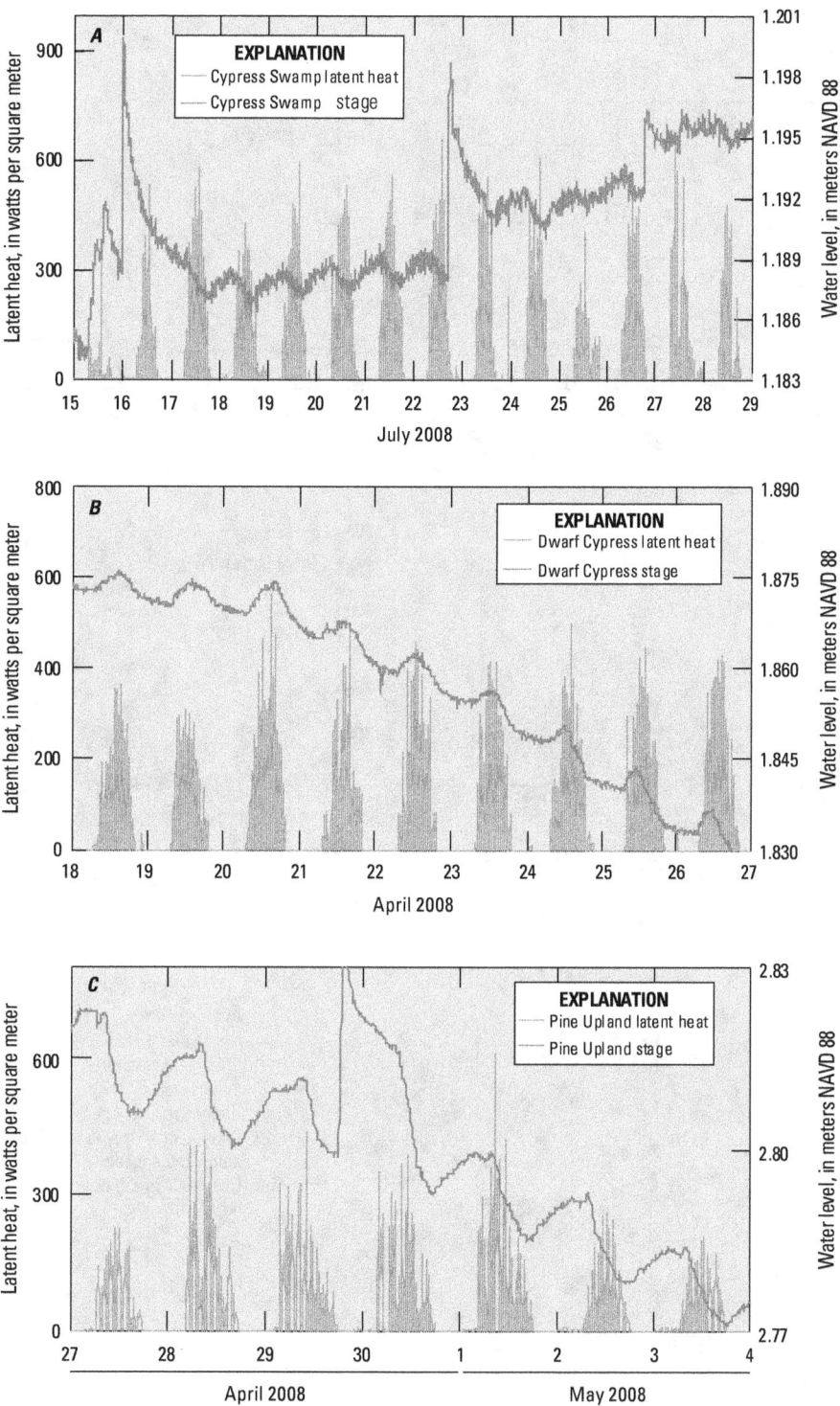

Figure 11. Water levels and latent-heat flux in response to evapotranspiration at the (*A*) Cypress Swamp, (*B*) Dwarf Cypress, and (*C*) Pine Upland sites in 2008. The Cypress Swamp and Dwarf Cypress sites are flooded and the Pine Upland site is dry.

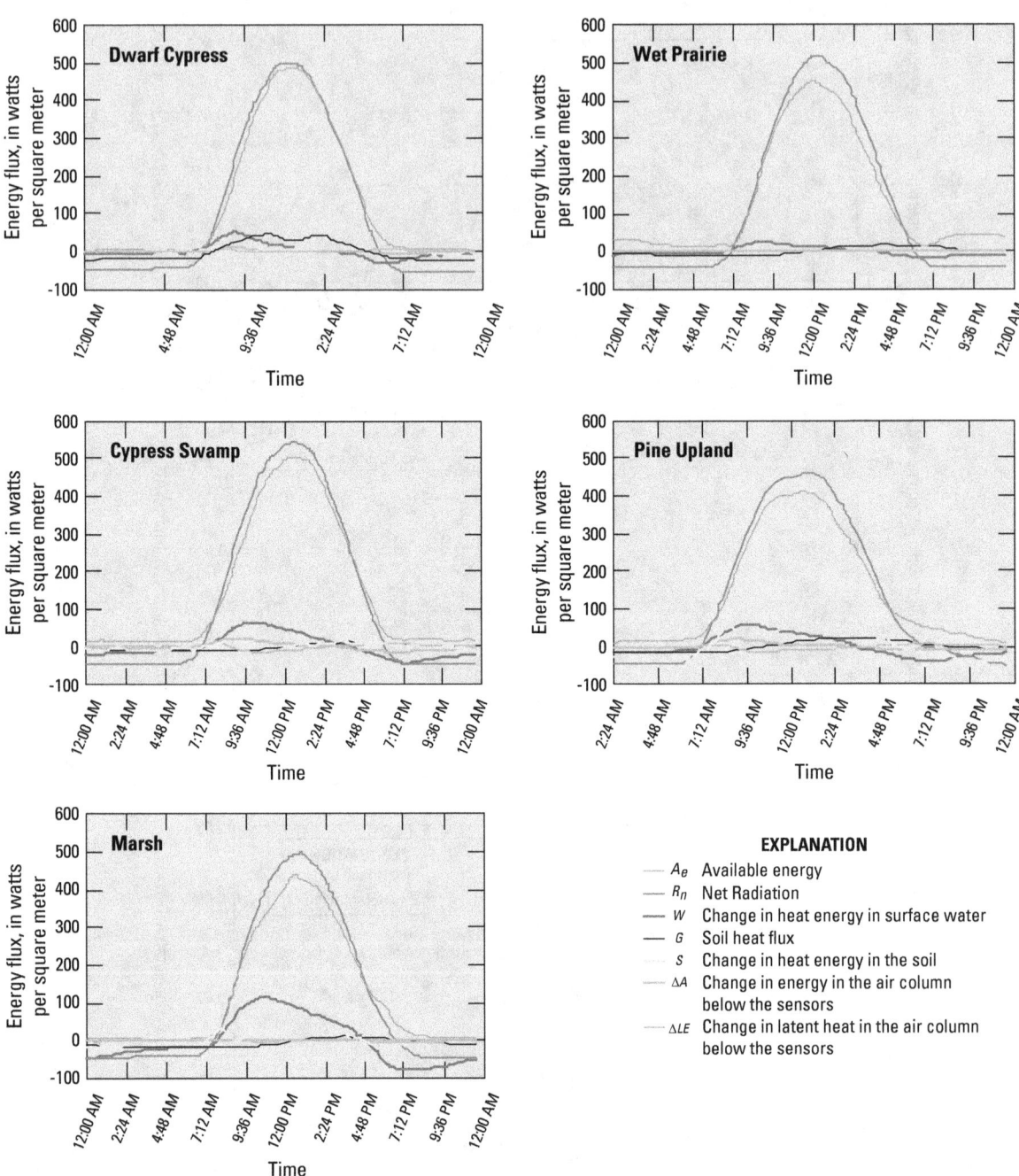

Figure 12. Surface-energy fluxes at the Dwarf Cypress, Wet Prairie, Cypress Swamp, Pine Upland, and Marsh evapotranspiration sites over a 24-hour period.

demonstrated that changes in heat energy in the surface water (W) can be a considerable component of subdaily and daily surface-energy budgets. Changes in latent heat (ΔLE) beneath the eddy covariance instrumentation were generally immaterial in all surface-energy budgets. Apparently, this surface-energy flux can be ignored in future analyses.

Seasonality in Evapotranspiration

Seasonality was apparent in the ET data at all five sites. ET was least from November to February when solar radiation was relatively small. A seasonality index was created to compare the seasonal variability of ET at the sites on an annual basis. The seasonality index (S_I) was computed as:

$$S_I = 1 - \frac{\overline{ET}_{min}}{\overline{ET}_{max}} \qquad (19)$$

where $\overline{ET}_{min}$ is the mean ET over 3 months with the least amount of ET, and $\overline{ET}_{max}$ is the mean ET over 3 months with the greatest amount of ET. Seasonality index values approaching 0.0 indicate little variability in monthly ET over a year. Index values close to 1.0 indicate considerable variability in monthly ET over a year at a specific site.

The seasonality index ranged from 0.33 to 0.63 at the five ET sites over the 3-year time period (table 5). In year 1, the seasonality index ranged from 0.47 to 0.63 and was greatest at the Cypress Swamp site and least at the Pine Upland site. In year 2, the seasonality index ranged from 0.38 to 0.56 and was greatest at the Pine Upland site and least at the Wet Prairie site. In year 3, the seasonality index ranged from 0.33 to 0.54 and was greatest at the Dwarf Cypress site and least at the Cypress Swamp site. Cypress trees are deciduous, which may partly explain relatively pronounced seasonality indexes. As previously mentioned, cypress trees appeared to reach full leaf-capacity in the early summer, while losing essentially all leaves in the early winter toward the end of hurricane season. Relatively large ET rates coincided with time periods when the cypress reached full leaf-capacity in the early summer.

Table 5. Seasonality index for evapotranspiration, October 10, 2007, to October 9, 2010.

[Year 1: October 10, 2007 to October 9, 2008; year 2: October 10, 2008 to October 9, 2009; year 3: October 10, 2009 to October 9, 2010]

Site	Year 1	Year 2	Year 3
Dwarf Cypress	0.56	0.47	0.54
Wet Prairie	.53	.38	.51
Cypress Swamp	.63	.50	.33
Pine Upland	.47	.56	.38
Marsh	.50	.42	.51

Spatial and Temporal Variability in Annual Evapotranspiration

Spatial and temporal variability in annual ET was characterized for future water budget analysis. Spatial variability in annual ET was considerable at the five ET sites (table 4). The maximum spatial differences in annual ET were about 400 mm in year 1 and about 300 mm in year 2. Annual ET rates ranged from about 800 mm at the Marsh site to about 1,200 mm at the Cypress Swamp site in year 1, and ranged from about 800 mm at the Marsh site to about 1,100 mm at the Cypress Swamp site in year 2. Annual ET rates were comparable at all five sites in year 3. In fact, ET differences between sites in year 3 may be within the experimental errors associated with the eddy covariance measurement technique. Note the Marsh site was recovering from extensive forest fire and drought conditions in 2007 as indicated by the increasing hydroperiod from year 1 to year 3 (table 4). These results suggest spatial differences in ET are likely reduced as ecosystems recover from fires and water-limiting drought conditions.

Temporal variability in annual ET was relatively small at the ET sites, except for the Marsh site which was recovering from fire and drought conditions. The maximum differences in annual ET were about 100 mm at the Dwarf Cypress site from year 1 to year 2, and about -100 mm at this same site from year 2 to year 3 (table 4). Thus, well-watered locations seem to have similar annual ET rates despite variations in the plant community.

Variations in Surface-Energy Fluxes during an Extreme Cold Front

As previously mentioned, Florida experienced an extreme cold front on about January 1, 2010. Minimum daily air temperatures of about 5 °C or less were measured at the ET sites every night for 2 weeks. This cold front had consequences for the ecology of the BCNP and subtropical southern Florida, including the death of many exotic species such as Blue Tilapia (*Oreochromis aureus*) and native species such as the Florida manatee and snook. The cold front presented a unique opportunity to examine how interaction of the surface-energy budget with the Bowen ratio determines latent-heat flux under extreme conditions. Noteworthy variations in surface-energy fluxes occurred as the ecosystem responded to more than 20 °C variations in cold-front air temperatures. No analysis was made for the Marsh site during the cold front because computer problems interfered with data compilations.

In general, as air temperature and net radiation increased at the Dwarf Cypress site during the cold front, surface water (W) absorbed a portion of the net radiation, diminishing energy available (A_e) for latent and sensible heat (fig. 13A). On January 9, 2010, from about 4:00 a.m to 8:30 AM, the air temperature dropped about 10 °C (fig. 13A). Net radiation (R_n) was negligible during this time period; however, energy was available (A_e) for latent- and sensible-heat fluxes

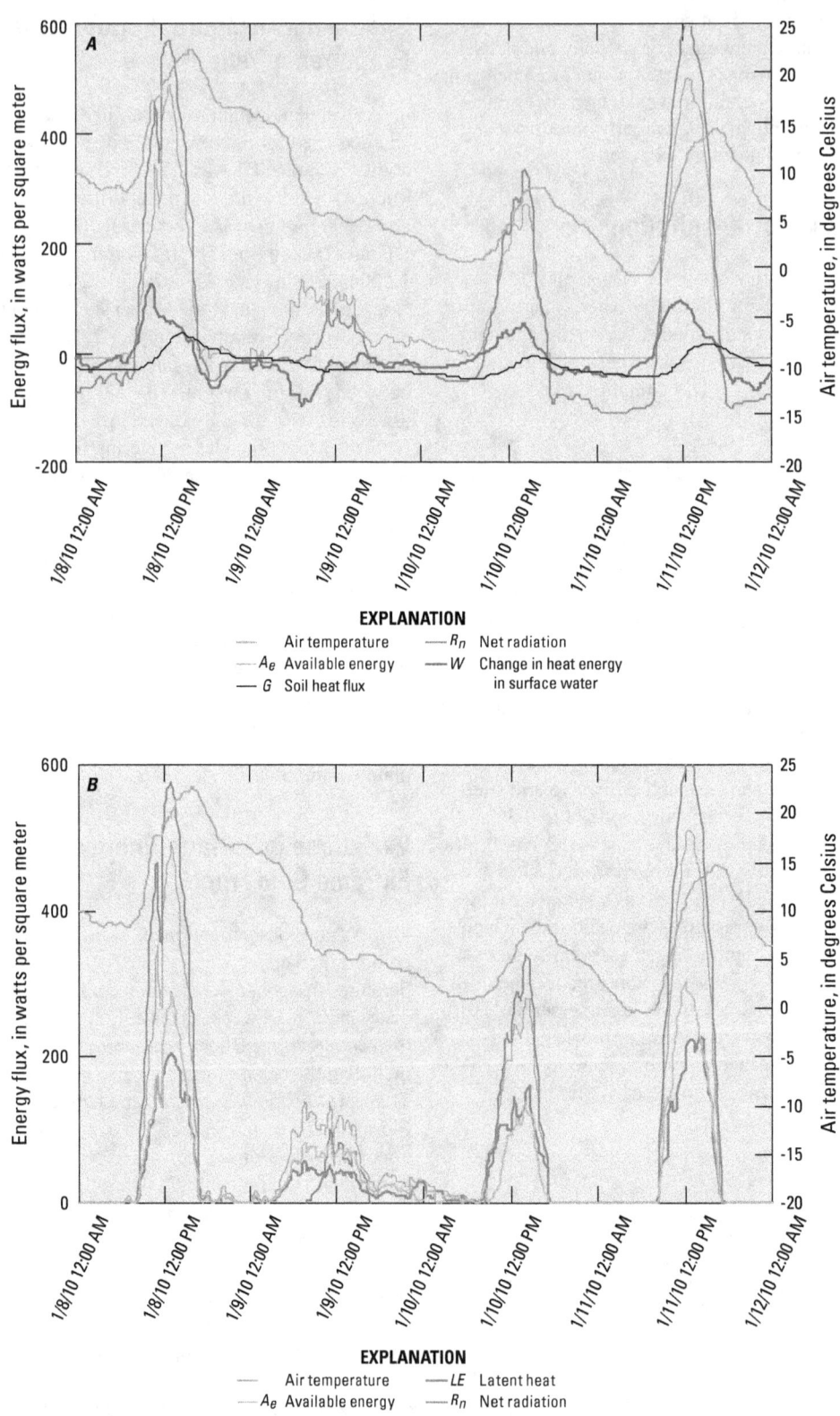

Figure 13. Air temperature in relation to (*A*) surface-energy fluxes and (*B*) latent- and sensible-heat fluxes at the Dwarf Cypress site during a cold front, January 8–12, 2010.

(fig. 13*B*). The Dwarf Cypress site was flooded with surface water, as indicated by the negative values for changes in heat energy stored in the surface water, *W* (fig. 13*A*). Decreasing air temperatures rapidly cooled the standing surface water, which made energy available for latent and sensible heat (eq. 1). At 10:30 AM, net radiation increased as the sun rose providing the energy for latent and sensible heat. During the same time period, changes in heat energy stored in the surface water (*W*) approached zero due to water temperatures equilibrating with air temperatures (fig. 13*A,B*). Thereafter, latent- and sensible-heat fluxes were mostly driven by net radiation for the remainder of the day. The Bowen ratio decreased over the remainder of the day (January 9, 2010) until latent heat approached sensible heat at about 2:00 PM.

Similar dynamics were observed at the Cypress Swamp site during the cold front. As air temperature and net radiation rose at the Cypress Swamp site, the surface water (*W*) absorbed a portion of net radiation making less energy available (*A_e*) for latent and sensible heat (fig. 14*A*). Changes in heat energy in the soil (*S*) and surface water (*W*) were of similar magnitude; however, changes in water-energy storage occurred more rapidly than changes in soil-energy storage due to the proximity of water to air (water is on top of the soil) and water's greater heat capacity. Positive latent- and sensible-heat fluxes occurred at night at the Cypress Swamp site during January 11-13, 2010, due to declining soil and water temperatures making energy available for latent and sensible heat. On January 10, 2010, the Bowen ratio increased over the course of the day as indicated by latent heat greater than sensible

heat in the morning, and sensible heat greater than latent heat in the afternoon (fig. 14*B*). The Bowen ratio remained such that sensible heat was greater than latent heat over the subsequent 4 days.

As air temperature and net radiation increased at the Pine Upland site during the cold front, the soil absorbed a portion of the rising net radiation making less energy available (*A_e*) for latent and sensible heat (fig. 15*A*). Because the site was dry with no surface water present to store heat-energy, the offset between net radiation and available energy was less at the Pine Upland site (fig. 15) than at the Dwarf Cypress and Cypress Swamp sites (figs. 13 and 14). Another consequence of the lack of surface water was less available energy and latent and sensible heat at night (fig. 15). During January 6-10, 2010, the Bowen ratio remained at values such that sensible heat was greater than latent heat during the day.

Similar dynamics were observed at the Wet Prairie site, which was dry during the cold front. As air temperature and net radiation rose, the soil absorbed a portion of net radiation making less energy available (*A_e*) for latent and sensible heat (fig. 16*A*). The offset between net radiation and available energy was less, because the Wet Prairie site was dry with no surface water present to store heat energy. Declining air temperature at night cooled the soil which released minor amounts of available energy for latent and sensible heat (fig. 16*B*). In contrast with the Pine Upland and Cypress Swamp sites, the Wet Prairie site maintained Bowen ratios such that latent heat was always greater than sensible heat during January 2-6, 2010 (fig. 16*B*).

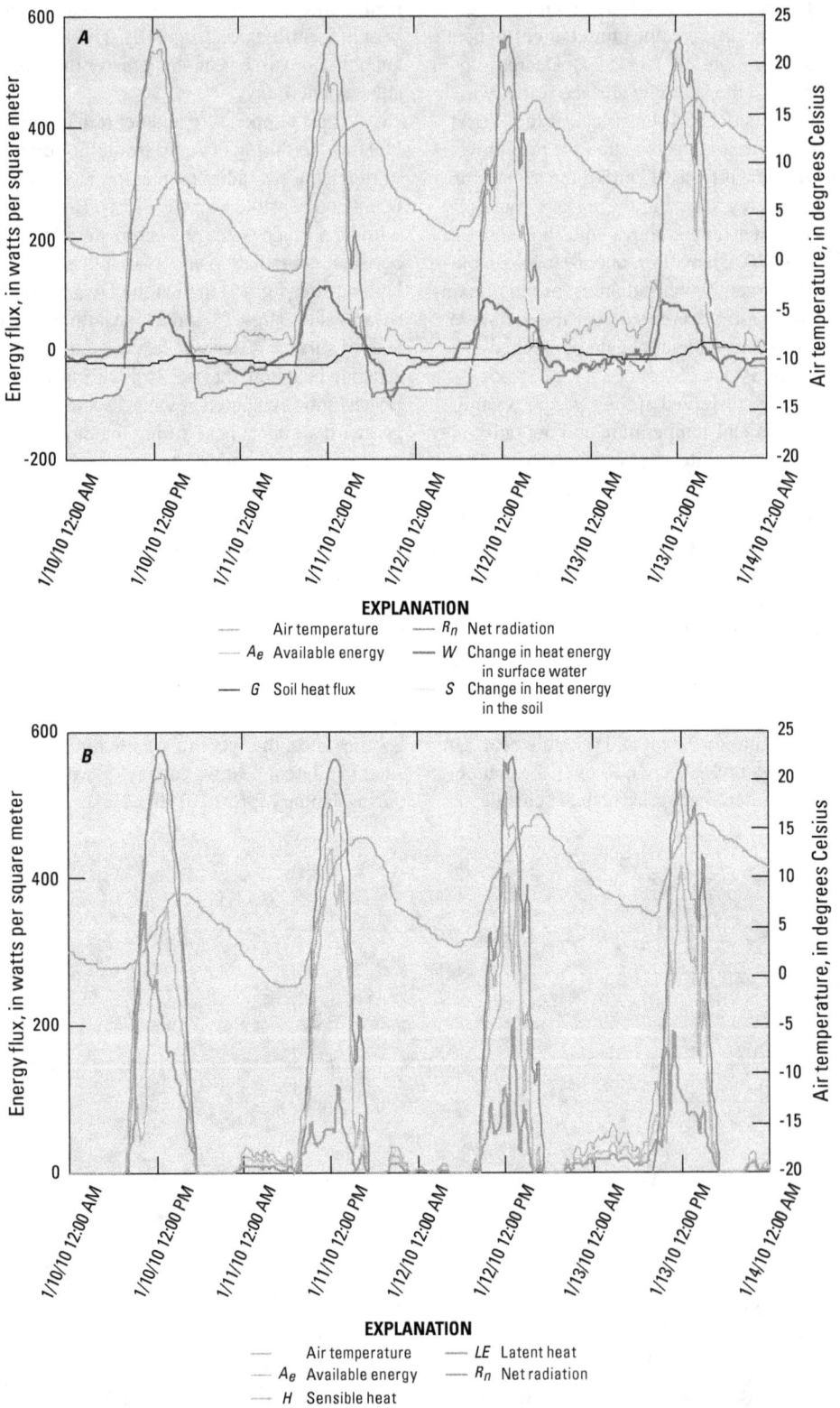

EXPLANATION

	Air temperature	R_n	Net radiation
	A_e Available energy	W	Change in heat energy in surface water
	G Soil heat flux	S	Change in heat energy in the soil

EXPLANATION

	Air temperature	LE	Latent heat
	A_e Available energy	R_n	Net radiation
	H Sensible heat		

Figure 14. Air temperature in relation to (*A*) surface-energy fluxes and (*B*) latent- and sensible-heat fluxes at the Cypress Swamp site during a cold front, January 10–14, 2010.

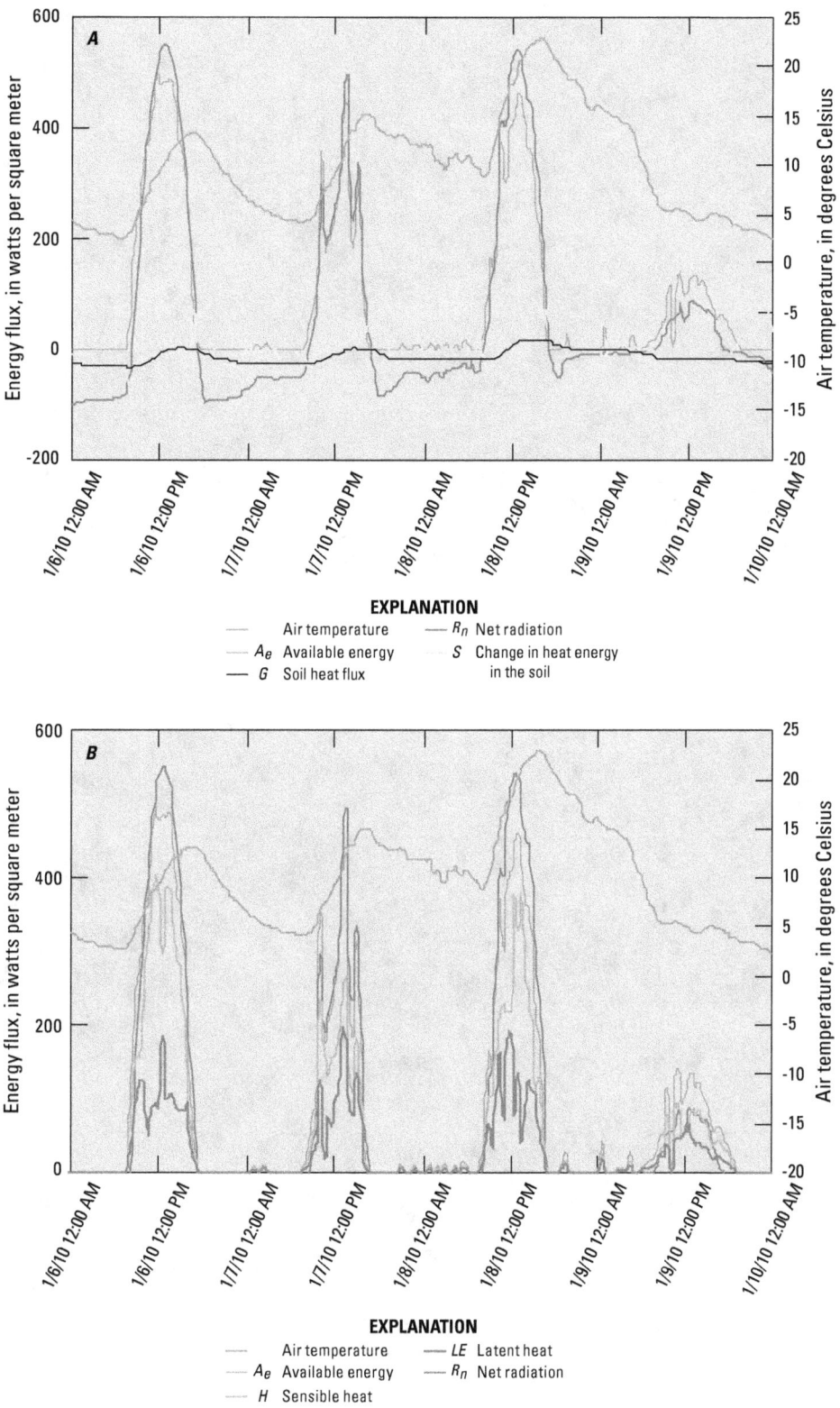

Figure 15. Air temperature in relation to (*A*) surface-energy fluxes and (*B*) latent- and sensible-heat fluxes at the Pine Upland site during a cold front, January 6–10, 2010.

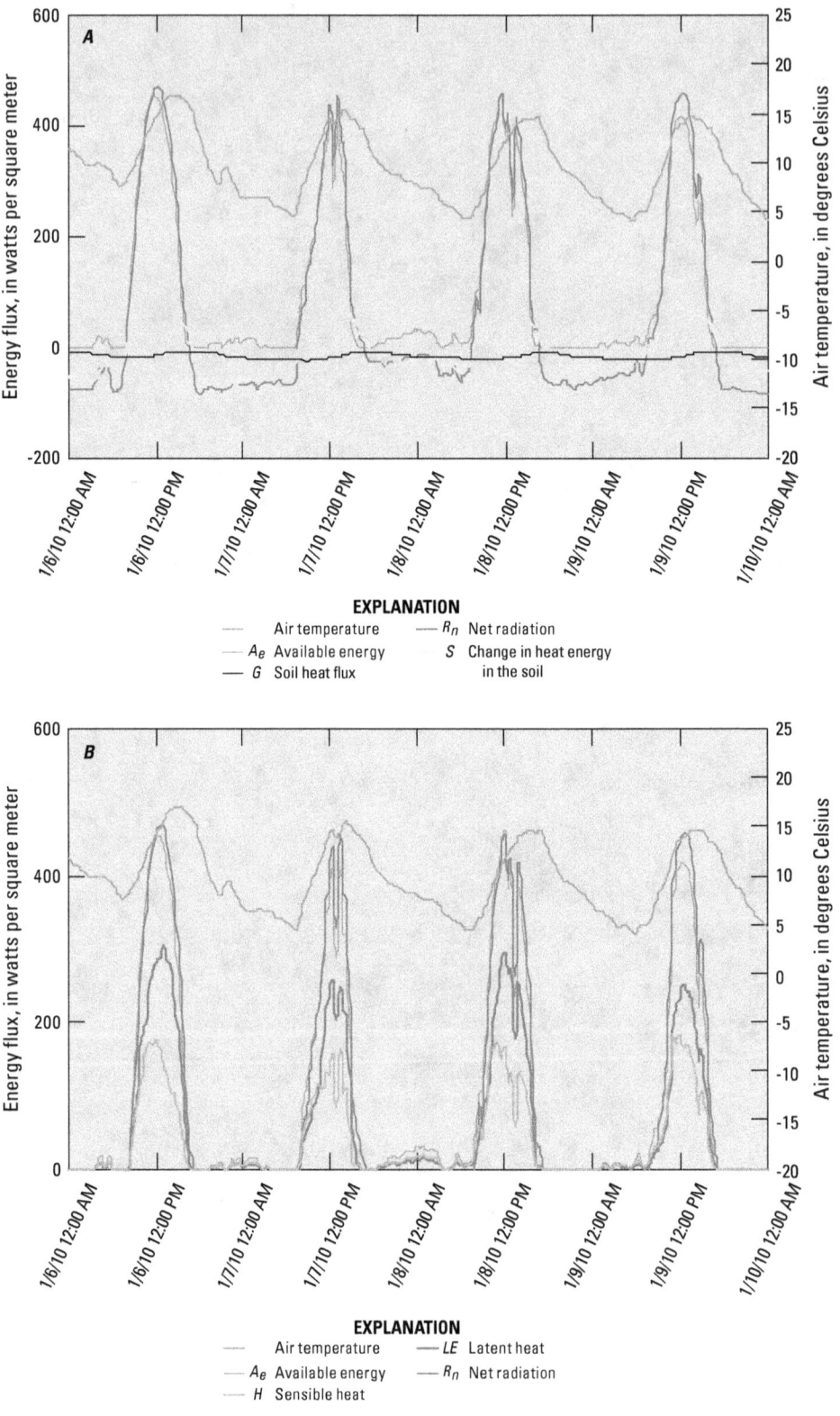

EXPLANATION

—	Air temperature	R_n	Net radiation
A_e	Available energy	S	Change in heat energy
G	Soil heat flux		in the soil

EXPLANATION

—	Air temperature	LE	Latent heat
A_e	Available energy	R_n	Net radiation
H	Sensible heat		

Figure 16.   Air temperature in relation to (*A*) surface-energy fluxes and
(*B*) latent- and sensible-heat fluxes at the Wet Prairie site during a cold
front, January 2–6, 2010.

Summary

Evapotranspiration (ET) was quantified over five spatially extensive plant communities in the Big Cypress National Preserve using the eddy covariance method between 2007 and 2010. Plant communities selected for study (Pine Upland, Wet Prairie, Marsh, Cypress Swamp, and Dwarf Cypress) are spatially extensive in southern Florida. Thus, the ET measurements collected as part of this investigation may provide insight to conditions in other humid subtropical regions including the Florida Everglades.

Characterization of spatial and temporal differences in ET and the surface-energy budget is challenging. Errors in eddy covariance ET measurements can be created by dirty hygrometer lenses, and/or misapplication of quality-assurance/quality-control procedures. These errors are not trivial and may sum to more than 100 percent of the initial flux measurement. Nevertheless, several spatial and temporal differences in ET and the surface-energy budget were determined that were consistent with field observations and understanding of the physical processes that govern ET variability.

Spatial differences in annual ET were considerable, apparently due to a combination of drought and forest fire at the Marsh site. In year 1 (October 10, 2007 to October 9, 2008), for example, the maximum spatial difference in annual ET was about 400 mm. Specifically, about 1,200 and 800 mm of ET were measured at the Cypress Swamp and Marsh sites, respectively. In year 2 (October 10, 2008 to October 9, 2009), the maximum spatial difference in annual ET was about 300 mm. More specifically, about 1,100 and 800 mm of ET were measured at the Cypress Swamp and Marsh sites, respectively. The Marsh site was recovering from extensive fire and drought conditions in 2007, as indicated by the increasing hydroperiod from year 1 to year 3. In year 3 (October 10, 2009 to October 9, 2010), annual ET rates were comparable at all five sites. In fact, spatial differences in annual ET between sites in year 3 may be within the experimental errors of the eddy covariance measurement technique.

Temporal variability in annual ET was relatively small at sites that were well watered (Dwarf Cypress, Wet Prairie, Cypress Swamp, and Pine Upland) over the 3-year period of record. Specifically, the maximum difference in annual ET from year 1 to year 2 was about 100 mm at the Dwarf Cypress site. The maximum difference in annual ET from year 2 to year 3 was about -100 mm also at the Dwarf Cypress site. Thus, locations that are well watered appear to have similar annual ET rates.

Diurnal water-level variability in response to ET was observed at all the ET sites. Diurnal water-level variability was less at flooded sites than at dry sites. Specifically, surface-water levels declined about 1.5 mm in response to ET at the flooded Cypress Swamp site during July 18-22, 2008. Surface-water levels declined about 10 mm in response to ET at the flooded Dwarf Cypress site during April 18-27, 2008. Diurnal water-level variability at the dry Pine Upland site was used with the concept of specific yield to verify the accuracy of eddy covariance ET estimates. Water levels repeatedly dropped about 15 mm on average in response to ET at the Pine Upland site from April 27 to May 4, 2008. ET was about 3 mm on each of these days, resulting in a reasonable estimate for specific yield of 0.2.

Seasonality was apparent in monthly ET and was generally greatest from March to October when solar radiation was relatively large, and least from November to February when solar radiation was relatively small. Monthly ET was greatest in the spring and summer at the Cypress Swamp site, reaching rates as large as 143 mm. The large ET rates at this site coincide with the most active period of cypress growth during late spring and early summer. Leaves on cypress trees begin to senesce in late summer, which apparently reduced transpiration.

A seasonality index can be used to compare the seasonal variability of ET at different sites on an annual basis. Seasonality index values close to 0.0 indicated little variability in monthly ET at a site over a year. Index values close to 1.0 indicated considerable variability in monthly ET at a site over a year. The seasonality index was greatest at the Cypress Swamp site in year 1 (equal to 0.63), and greatest at the Pine Upland site in year 2 (equal to 0.56). In year 3, the seasonality index at the Dwarf Cypress site was 0.54. Cypress trees are deciduous, which may partly explain relatively pronounced seasonality indexes. This seasonality index may be useful as a starting point for comparisons of ET estimated in future studies with the ET values reported herein.

Trends were observed in the surface-energy budget. For example, available energy was generally less than net radiation until early to late afternoon, as a portion of net radiation was used to heat the soil, surface-water, and air column. Available energy was generally greater than net radiation in the late afternoon and night as the energy used to heat the soil, surface water, and air column was released back into the atmosphere, and thus, made available for latent and sensible heat. These results suggest relatively mild amounts of ET can occur well into the night as landscapes radiate heat energy away from land surface.

Net radiation and available energy explained most of the variability in ET observed at the monitoring sites. Net radiation was generally greatest at the Cypress Swamp site, averaging about 130 W/m² over the 3-year period of record. The Cypress Swamp site apparently has the smallest albedo, which likely is due to the relatively dark, densely spaced, and tall cypress trees. Net radiation was generally least at the Dwarf Cypress site, averaging about 115 W/m² over 3 years. The Dwarf Cypress site apparently has the largest albedo, which likely is due to the sparse canopy and a highly reflective, calcareous, periphyton-covered land surface. Furthermore, mean annual net radiation was least in the first year of the study, which likely was due to greater cloud cover during a relatively wet year. In contrast, net radiation was greatest in the second year of the study, which likely was due to less cloud cover during a relatively dry year.

Florida experienced an extreme cold front around January 1, 2010. Air temperatures of about 5 degrees Celsius or less were measured at the sites for several weeks. This cold front had consequences for the ecology of the BCNP and subtropical south Florida, including the death of many exotic species such as the Blue Tilapia (Oreochromis aureus) and native species such as the Florida manatee and snook. The cold front presented a unique opportunity to examine how interaction of the surface-energy budget with the Bowen ratio determines latent-heat flux under cold conditions. Curious variations in surface-energy fluxes occurred as the ecosystem responded to over 20-degree- Celsius variations in cold-front air temperatures.

Available energy increased as air temperature decreased, especially at sites near standing surface water. In fact, declining air temperatures at sites near standing water apparently created enough energy for occurrence of ET at night. In contrast, available energy decreased as air temperature increased, especially at sites with standing surface water. The surface water absorbed a portion of net radiation making less energy available for latent and sensible heat. Changes in heat energy in the soil and surface water were of similar magnitude during the cold front; however, changes in water-energy storage occurred more rapidly than changes in soil-energy storage due to the proximity of water to the atmosphere (water on top of soil).

Available water is computed as the difference between rainfall and ET on an annual and monthly basis. Available water was always positive on an annual basis, indicating surplus rainfall was always available for aquifer recharge and runoff toward the coast. Monthly available water ranged from -126 to 405 mm at the ET sites from October 2007 to September 2010. The Pine Upland and Marsh sites received the greatest amount of monthly available water (about 400 mm each) in August 2008. The least amount of available water (-126 mm) was measured at the Cypress Swamp site in April 2009. Negative available water generally occurred at all the sites in the relatively dry winter (October to May). Positive available water generally occurred at all the sites in the hot and humid summer (June to September). Available water was concentrated within 3 to 4 months (June to September) over a year. Variations in monthly available water were mostly explained by rainfall variability.

References Cited

Abtew, W., 1996a, Evapotranspiration measurements and modeling for three wetland systems: Journal of American Water Resources Association, no. 32, p. 465–473.

Abtew, W., 1996b, Lysimeter study of evapotranspiration from a wetland; in C.R. Camp, E.J. Sadler, and R.E. Yoder (eds.) Evapotranspiration and Irrigation Scheduling: Proceedings of the International Conference, San Antonio TX, USA.

Abtew, W., and Obeysekera, J., 1995, Lysimeter study of evapotranspiration of cattails and comparison of three estimation methods: American Society of Civil Engineers, no. 38, p. 121–129.

Baldocchi, D.D., Hicks, B.B., and Meyers, T.P., 1988, Measuring biosphere-atmosphere exchanges of biologically related gases with micrometeorological methods: Ecology, v. 69, no. 5, p. 1331-1340.

Bidlake, W.R., Woodham, W.M., and Lopez, M.A., 1996, Evapotranspiration from areas of native vegetation in west-central Florida: U.S. Geological Survey Water-Supply Paper 2430, 35 p.

Bowen, I.S., 1926, The ratio of heat losses by conduction and by evaporation from any water surface: Physical Review, 2nd series, v. 27, no. 6, p. 779-787.

Burba, G., and Anderson, D., 2007, Introduction to the eddy covariance method: General guidelines and conventional workflow: LI-COR Biosciences, 141 p.

Campbell, G.S., and Norman, J.M., 1998, An introduction to environmental biophysics: New York, Springer, 286 p.

Campbell Scientific, Inc., 1990, TCAV averaging soil thermocouple probe instruction manual: Logan, Utah, 2 p.

Duever, M.J, Carlson, J.E. Meeder, J.F., Duever, L.C., Gunderson, L.H. Riopelle, L.A., Alexander, T.R., Myers, R.L., and Spangler, D.P., 1986, The Big Cypress National Preserve: New York, National Audubon Society, 455 p.

Dyer, A.J., 1961, Measurements of evaporation and heat transfer in the lower atmosphere by an automatic eddy covariance technique: Quarterly Journal of the Royal Meteorological Society, v. 87, p. 401-412.

Foken, T., 2008, The energy balance closure problem: Ecological Applications, v., 18, no. 6, p. 1351–1367

German, E.R., 2000, Regional evaluation of evapotranspiration in the Everglades: U.S. Geological Survey Water Resources Investigations Report 00–4217, 48 p.

Goulden, M.L., Munger, J.W., Fan, S-M, Daube, B.C., and Wofsy, S.C., 1996, Measurements of carbon sequestration by long-term eddy covariance: Methods and a critical evaluation of accuracy: Global Change Biology, v. 2, p. 169-182.

Islam, S., Eltahir, E., and Jiang, L., 2002, Satellite-based evapotranspiration estimation: Final report prepared for the South Florida Water Management District, West Palm Beach, FL, USA.

Jacobs, J., Mecikalski, J., and Paech, S., 2008, Satellite-based solar radiation, net radiation, and potential and reference evapotranspiration estimates over Florida: Report prepared for the South Florida Water Management District, 138 p.

Kaimal, J.C., and Gaynor, J.E., 1991, Another look at sonic thermometry: Boundary-Layer Meteorology, v. 56, p. 401-410.

Knowles, L., Jr., 1996, Estimation of evapotranspiration in the Rainbow Springs and Silver Springs basins in north-central Florida: U.S. Geological Survey Water-Resources Investigations Report 96-4024, 37 p.

Lee, X., and Black, T.A., 1993, Atmospheric turbulence within and above a Douglas-fir stand. Part II: Eddy fluxes of sensible heat and water vapor: Boundary-Layer Meteorology, v. 64, p. 369-389.

McPherson, B.F., 1973, Vegetation map of southern parts of subareas A and C, Big Cypress Swamp, Florida: U.S. Geological Survey Hydrologic Atlas HA-492.

Merritt, M.L., 1996, Numerical simulation of a plume of brackish water in the Biscayne aquifer originating from a flowing artesian well, Dade County, Florida: U.S. Geological Survey Water-Supply Paper 2464, 74 p.

Monteith, J. L., 1965, Evaporation and environment; in Proceedings of the 19th Symposium of the Society for Experimental Biology: New York, Cambridge University Press, p. 205-233.

Monteith, J.L., and Unsworth, M.H., 1990, Principles of environmental physics: London, Edward Arnold, 291 p.

Moore, C.J., 1976, Eddy flux measurements above a pine forest: Quarterly Journal of the Royal Meteorological Society, v. 102, p. 913-918.

Nash, J.E., and Sutcliffe, J.V., 1970, River flow forecasting through conceptual models, part I, A discussion of principles: Journal of Hydrology, v. 10, no. 3, p. 282-290.

Price, R.M., Nuttle, W.K., Cosby, B.J., and Swart, P.K., 2007, Variation and uncertainty in evaporation from a subtropical estuary: Florida Bay, Estuaries and Coasts, v. 30, no. 3, p. 497-506

Priestley, C.H.B., and Taylor, R.J., 1972, On the assessment of surface heat flux and evaporation using large scale parameters: Monthly Weather Review, no. 100, p. 81–92.

Reese, R.S., and Cunningham, K.J., 2000, Hydrogeology of the gray limestone aquifer in southern Florida: U.S. Geological Survey Water-Resources Investigations Report 99-4213, 244 p.

Rutchey, K., Schall, T.N., Doren, R.F., Atkinson, A., and others, 2006, Vegetation classification for south Florida natural areas: U.S. Geological Survey Open-File Report 2006-1240, 142 p.

Schotanus, P., Nieuwstadt, F.T.M., and de Bruin, H.A.R., 1983, Temperature measurement with a sonic anemometer and its application to heat and moisture fluxes: Boundary-Layer Meteorology, v. 50, p. 81-93.

Schuepp, P.H., Leclerc, M.Y., MacPherson, J.I., and Desjardins, R.L., 1990, Footprint prediction of scalar fluxes from analytical solutions of the diffusion equation: Boundary-Layer Meteorology, v. 50, p. 355-373.

Shoemaker, W.B., 1998, Geophysical delineation of hydrostratigraphy in the Big Cypress National Preserve, Florida: Tampa MS Thesis, University of South Florida, 119 p.

Shoemaker, W.B., Huddleston, S., Boudreau, C.L., and O'Reilly, A.M., 2008, Sensitivity of wetland saturated hydraulic heads and water budgets to evapotranspiration: Wetlands, v. 28, no. 4, p. 1040–1047.

Shoemaker, W.B., and Sumner, D.M., 2005, Estimating changes in heat energy stored within a column of wetland surface water and factors controlling their importance in the surface energy budget: Water Resources Research, no. 41:W10411.

Shoemaker, W.B., and Sumner, D.M., 2006, Alternate corrections for estimating actual wetland evapotranspiration from potential evapotranspiration: Wetlands, v. 26, no. 2, p. 528–543.

Stannard, D.I., 1993, Comparison of Penman-Monteith, Shuttleworth-Wallace, and modified Priestley-Taylor evapotranspiration models for wildland vegetation in semiarid rangeland: Water Resources Research, no. 29, p. 1379–1392.

Stull, R.B., 1988, An introduction to boundary layer meteorology: Boston, Kluwer Academic Publishers, 666 p.

Sumner, D.M., 1996, Evapotranspiration from successional vegetation in a deforested area of the Lake Wales Ridge, Florida: U.S. Geological Survey Water- Resources Investigations Report 96-4244, 38 p.

Sumner, D. M., 2001, Evapotranspiration from a cypress and pine forest subjected to natural fires in Volusia County, Florida, 1998–99: U.S. Geological Survey Water-Resources Investigations Report 01-4245, 56 p.

Sumner, D.M., and Belaineh, G., 2006, Evaporation, precipitation, and associated salinity changes at a humid, subtropical estuary: Estuaries, no. 28., p. 844–55.

Swancar, A., Lee, T.M., and O'Hare, T.M., 2000, Hydrogeologic setting, water budget, and preliminary analysis of ground-water exchange at Lake Starr, a seepage lake in Polk County, Florida: U.S. Geological Survey Water-Resources Investigations Report 00- 4030, 72 p.

Tanner, B.D., and Greene, J.P., 1989, Measurement of sensible heat and water vapor fluxes using eddy covariance methods: Final report prepared for U.S. Army Dugway Proving Grounds, Dugway, Utah, 17 p.

Tanner, B.D., Swiatek, E., and Greene, J.P., 1993, Density fluctuations and use of the krypton hygrometer in surface flux measurements: Management of irrigation and drainage systems: Irrigation and Drainage Division, American Society of Civil Engineers, July 21-23, 1993, Park City, Utah, p. 945-952.

Tanner, C.B., and Thurtell, G.W., 1969, Anemoclinometer measurements of Reynolds stress and heat transport in then atmospheric surface layer: University of Wisconsin Technical Report ECOM-66- G22-F, 82 p.

Wetzel, P.R., Sklar, F.H., Coronado, C.A., Troxler, T.G., Krupa, S.L., Sullivan, P.L., Ewe, S., and Orem, W.H., 2011, Biogeochemical processes on tree islands in the greater Everglades: Initiating a new paradigm: Critical Reviews in Environmental Science and Technology, no. 41, supp. 1, p. 670-701.

Appendix 1.   Monthly Water- and Energy-Balance Calculations at the Evapotranspiration Sites, March 2007 to September 2010

Date	Total evapotranspiration, in millimeters				
	Dwarf Cypress	**Wet Prairie**	**Cypress Swamp**	**Pine Upland**	**Marsh**
March-07					
April-07					
May-07	110		141	78	
June-07	104		126	97	
July-07	99		130	99	76
August-07	102		126	110	92
September-07	83		98	87	73
October-07	62	52	77	67	68
November-07	53	67	53	57	49
December-07	46	56	48	48	42
January-08	49	53	49	42	39
February-08	58	58	59	49	49
March-08	79	74	86	65	65
April-08	111	96	129	86	78
May-08	109	124	161	94	82
June-08	95	96	149	90	80
July-08	105	100	139	101	87
August-08	112	109	131	93	82
September-08	105	112	117	88	93
October-08	92	96	92	73	70
November-08	72	73	61	53	46
December-08	58	65	59	49	40
January-09	61	68	62	50	43
February-09	66	62	75	50	48
March-09	79	71	104	67	64
April-09	85	86	146	73	78
May-09	94	96	131	83	75
June-09	112	103	130	101	50
July-09	124	104	132	107	93
August-09	128	109	124	113	102
September-09	102	88	107	93	87
October-09	107	67	97	84	91
November-09	74	69	67	60	61
December-09	102	59	55	49	66
January-10	57	57	51	45	55
February-10	60	66	52	46	60
March-10	83	90	79	62	90
April-10	91	95	92	88	105
May-10	94	119	132	116	107
June-10	93	122	131	130	96
July-10	89	118	125	115	122
August-10	85	117	114	116	114
September-10	70	88	97	92	103

Date	Rainfall, in millimeters				
	Dwarf Cypress	**Wet Prairie**	**Cypress Swamp**	**Pine Upland**	**Marsh**
March-07	88	88	64	13	11
April-07	94	94	149	144	59
May-07	81	81	83	54	174
June-07	314	314	422	214	227
July-07	220	220	308	234	229
August-07	185	185	237	137	182
September-07	247	247	245	128	269
October-07	75	75	62	76	88
Nove mber-07	4	4	9	13	8
December-07	18	18	70	19	18
January-08	15	15	8	21	19
February-08	134	134	131	144	126
March-08	32	32	38	54	67
April-08	55	55	47	57	54
May-08	43	43	21	19	24
June-08	256	256	215	260	354
July-08	245	245	344	230	210
August-08	386	386	423	487	474
September-08	172	172	234	139	170
October-08	99	99	70	106	112
November-08	21	21	7	5	7
December-08	17	17	12	19	14
January-09	5	5	4	6	4
February-09	20	20	20	18	12
March-09	32	32	38	25	4
April-09	7	7	9	6	9
May-09	274	274	142	195	276
June-09	300	300	310	351	333
July-09	161	161	214	136	145
August-09	124	124	267	145	225
September-09	157	157	105	150	254
October-09	78	78	49	30	47
November-09	61	61	48	42	29
December-09	55	55	53	101	80
January-10	44	44	111	42	60
February-10	73	73	40	39	89
March-10	56	56	66	120	151
April-10	128	128	185	86	91
May-10	137	137	59	168	97
June-10	136	136	138	124	128
July-10	132	132	156	113	150
August-10	269	269	354	250	280
September-10	138	138	145	114	138

Date	Available water, in millimeters				
	Dwarf Cypress	**Wet Prairie**	**Cypress Swamp**	**Pine Upland**	**Marsh**
March-07					
April-07					
May-07	-28		-58	-24	
June-07	211		296	118	
July-07	120		178	135	152
August-07	82		111	27	90
September-07	164		147	41	195
October-07	13	23	-15	9	20
November-07	-49	-63	-44	-45	-40
December-07	-28	-39	22	-29	-24
January-08	-34	-38	-41	-21	-20
February-08	75	75	73	94	77
March-08	-47	-42	-48	-11	2
April-08	-56	-40	-82	-29	-24
May-08	-67	-81	-140	-75	-58
June-08	161	160	67	169	273
July-08	141	145	206	129	122
August-08	274	277	292	394	392
September-08	67	59	118	51	77
October-08	6	3	-23	33	42
November-08	-52	-52	-54	-48	-39
December-08	-41	-48	-46	-30	-26
January-09	-56	-64	-58	-44	-39
February-09	-46	-42	-55	-32	-36
March-09	-48	-39	-66	-43	-60
April-09	-78	-78	-136	-67	-69
May-09	180	178	11	113	200
June-09	188	198	180	250	283
July-09	36	57	81	29	51
August-09	-3	15	143	32	122
September-09	55	69	-2	57	167
October-09	-29	11	-48	-54	-44
November-09	-13	-7	-19	-18	-31
December-09	-46	-3	-2	52	14
January-10	-13	-13	60	-3	5
February-10	13	7	-11	-8	29
March-10	-27	-34	-13	57	61
April-10	37	32	93	-3	-14
May-10	44	18	-72	52	-10
June-10	43	14	7	-6	32
July-10	43	13	30	-3	28
August-10	185	152	240	134	166
September-10	68	51	48	21	35

| Date | Mean volumetric water content (VWC) ratio of saturated to total porosity | | | | |
	Dwarf Cypress	Wet Prairie	Cypress Swamp	Pine Upland	Marsh
April-07		0.66		0.07	
May-07		0.66		0.07	
June-07		0.64		0.12	
July-07		0.63		0.35	
August-07		0.63		0.33	
September-07		0.62		0.31	
October-07		0.63		0.3	
November-07		0.62		0.32	
December-07		0.63		0.32	
January-08		0.65	0.57	0.26	
February-08		0.64	0.59	0.27	
March-08		0.53	0.6	0.35	
April-08		0.55	0.63	0.34	
May-08		0.67	0.54	0.19	
June-08		0.68	0.52	0.17	
July-08		0.67	0.62	0.33	
August-08		0.67	0.6	0.34	
September-08		0.67	0.58	0.32	
October-08		0.68	0.58	0.3	
November-08		0.7	0.58	0.29	
December-08		0.7	0.6	0.3	
January-09		0.65	0.6	0.31	
February-09		0.6	0.57	0.28	
March-09		0.49	0.51	0.12	
April-09		0.46	0.34	0.06	
May-09		0.68	0.13	0.04	
June-09		0.68	0.5	0.31	
July-09		0.7	0.54	0.31	
August-09		0.71	0.54	0.31	
September-09		0.71	0.54	0.31	
October-09		0.67	0.55	0.32	
November-09		0.71	0.57	0.33	
December-09		0.71	0.58	0.36	
January-10		0.68	0.59	0.37	
February-10		0.67	0.59	0.37	
March-10		0.62	0.59	0.37	
April-10		0.63	0.59	0.37	
May-10		0.59	0.58	0.37	
June-10		0.61	0.56	0.37	
July-10		0.63	0.61	0.37	
August-10		0.7	0.61	0.37	
September-10		0.72	0.58	0.36	

Date	Mean net radiation, in watts per square meter				
	Dwarf Cypress	Wet Prairie	Cypress Swamp	Pine Upland	Marsh
March-07					
April-07	172		152	141	
May-07	159	149	169	132	
June-07	155	160	160	151	132
July-07	142	142	158	140	123
August-07	144	144	149	150	146
September-07	125	124	128	127	121
October-07	96	93	103	93	94
November-07	83	83	86	83	76
December-07	65	67	72	68	62
January-08	72	70	80	68	65
February-08	93	92	97	89	88
March-08	124	119	130	114	115
April-08	143	151	159	150	149
May-08	168	169	195	176	161
June-08	132	136	152	138	142
July-08	128	135	134	142	128
August-08	129	135	125	124	131
September-08	127	129	124	126	118
October-08	103	120	103	100	80
November-08	83	85	91	84	70
December-08	62	66	72	62	79
January-09	69	75	81	71	108
February-09	100	104	114	103	120
March-09	113	120	125	115	150
April-09	156	163	168	152	138
May-09	145	144	162	179	159
June-09	134	156	148	160	147
July-09	161	156	152	174	160
August-09	173	161	151	141	156
September-09	133	133	120	120	133
October-09	119	124	123	117	129
November-09	78	79	85	77	80
December-09	60	60	63	58	64
January-10	63	62	109	60	68
February-10	88	85	89	90	93
March-10	111	112	117	109	117
April-10	149	141	144	132	138
May-10	158	162	168	162	156
June-10	161	168	162	172	151
July-10	151	156	154	151	153
August-10	131	131	124	140	136
September-10	117	123	120	116	129

Date	Mean air temperature, in degrees Celsius				
	Dwarf Cypress	Wet Prairie	Cypress Swamp	Pine Upland	Marsh
March-07					
April-07	27			24	
May-07	27		23	24	
June-07	26	24	24	26	26
July-07	27	26	26	27	26
August-07	28	27	27	28	27
September-07	27	29	28	27	26
October-07	27	28	27	26	26
November-07	22	26	26	22	20
December-07	22	22	22	22	20
January-08	19	21	22	19	18
February-08	22	18	19	22	20
March-08	22	21	22	22	20
April-08	23	21	22	23	22
May-08	25	22	23	25	24
June-08	27	25	25	26	25
July-08	27	26	26	27	24
August-08	28	27	27	28	27
September-08	28	28	28	28	27
October-08	25	28	28	25	26
November-08	21	27	25	20	16
December-08	21	20	20	20	19
January-09	18	17	20	18	16
February-09	18	17	18	18	17
March-09	20	20	18	20	19
April-09	23	22	20	23	22
May-09	23	24	23	24	24
June-09	27	27	24	27	27
July-09		27	28	28	27
August-09		28	28	28	28
September-09		27	27	28	27
October-09	27	27	26	27	26
November-09	23	22	23	23	21
December-09	21	20	21	21	0
January-10	16	15	20	16	15
February-10	16	16	16	16	16
March-10	18	17	18	18	20
April-10	22	22	22	24	22
May-10	26	26	26	26	27
June-10	28	28	28	28	27
July-10	28	28	28	28	28
August-10	28	27	28	28	28
September-10	28	27	27	27	27

Date	Latent heat, in watts per square meter				
	Dwarf Cypress	**Wet Prairie**	**Cypress Swamp**	**Pine Upland**	**Marsh**
March-07					
April-07	125		125	80	
May-07	100		129	71	
June-07	98		118	91	85
July-07	90		118	91	70
August-07	93		115	100	84
September-07	78		92	82	69
October-07	56	67	70	61	61
November-07	53	64	50	54	46
December-07	42	52	44	44	39
January-08	45	49	45	39	36
February-08	57	57	57	48	48
March-08	72	68	79	59	59
April-08	99	91	110	80	74
May-08	100	113	147	86	75
June-08	89	90	140	85	79
July-08	95	91	126	92	80
August-08	101	99	119	85	74
September-08	99	106	110	83	88
October-08	84	85	84	67	64
November-08	68	69	57	50	44
December-08	53	59	54	44	37
January-09	56	63	56	45	39
February-09	67	64	76	50	49
March-09	73	65	95	61	59
April-09	85	81	131	69	73
May-09	84	88	99	75	69
June-09	104	97	102	95	88
July-09	113	94	99	97	85
August-09	117	99	99	103	93
September-09	96	83	84	87	82
October-09	97	94	78	76	83
November-09	70	65	58	57	57
December-09	49	54	42	44	61
January-10	48	53	40	42	50
February-10	58	67	45	47	62
March-10	73	82	57	57	82
April-10	99	90	81	78	99
May-10	85	109	120	106	97
June-10	87	116	123	122	90
July-10	81	108	114	105	111
August-10	77	108	104	105	104
September-10	66	83	91	87	97

Date	Sensible heat, in watts per square meter				
	Dwarf Cypress	**Wet Prairie**	**Cypress Swamp**	**Pine Upland**	**Marsh**
March-07					
April-07	69		50	88	
May-07	71		59	80	
June-07	67		51	77	87
July-07	64		49	68	71
August-07	64		43	72	82
September-07	62		43	65	72
October-07	49	41	40	53	54
November-07	45	45	42	56	56
December-07	37	38	38	44	47
January-08	43	43	50	44	51
February-08	47	48	51	51	59
March-08	61	60	57	58	70
April-08	69	68	61	73	86
May-08	87	70	82	93	93
June-08	63	53	64	73	78
July-08	59	51	63	71	81
August-08	60	48	64	68	68
September-08	59	43	59	70	61
October-08	52	46	60	61	53
November-08	51	47	71	66	45
December-08	43	38	57	51	33
January-09	48	43	65	59	40
February-09	63	62	88	79	58
March-09	64	72	86	79	62
April-09	79	101	88	102	77
May-09	67	67	70	76	70
June-09	64	69	70	76	69
July-09	74	65	61	80	80
August-09	68	67	64	77	80
September-09	73	58	58	75	58
October-09	55	65	64	68	59
November-09	42	35	55	58	46
December-09	45	30	46	47	58
January-10	46	33	58	58	77
February-10	76	45	69	71	71
March-10	76	59	71	75	88
April-10	76	55	68	70	93
May-10	77	61	62	77	88
June-10	79	63	57	76	68
July-10	76	59	54	67	50
August-10	80	51	53	70	50
September-10	79	52	51	62	41

Date	Bowen ratio, unitless				
	Dwarf Cypress	Wet Prairie	Cypress Swamp	Pine Upland	Marsh
March-07					
April-07	0.55		0.4	1.09	
May-07	0.71		0.46	1.13	
June-07	0.68		0.43	0.85	1.02
July-07	0.71		0.42	0.75	1.02
August-07	0.69		0.37	0.72	0.97
September-07	0.79		0.47	0.8	1.05
October-07	0.88	0.61	0.57	0.86	0.88
November-07	0.85	0.71	0.83	1.04	1.22
December-07	0.88	0.74	0.86	1	1.21
January-08	0.97	0.89	1.13	1.13	1.43
February-08	0.82	0.83	0.9	1.07	1.23
March-08	0.84	0.89	0.72	0.98	1.19
April-08	0.69	0.75	0.55	0.92	1.17
May-08	0.88	0.61	0.56	1.08	1.24
June-08	0.71	0.59	0.46	0.86	0.99
July-08	0.62	0.56	0.5	0.77	1.01
August-08	0.59	0.49	0.54	0.8	0.92
September-08	0.59	0.41	0.54	0.85	0.7
October-08	0.62	0.54	0.72	0.91	0.83
November-08	0.75	0.68	1.23	1.32	1.03
December-08	0.81	0.64	1.06	1.15	0.9
January-09	0.87	0.68	1.14	1.29	1.01
February-09	0.94	0.97	1.16	1.56	1.18
March-09	0.88	1.1	0.91	1.29	1.06
April-09	0.93	1.24	0.67	1.48	1.05
May-09	0.8	0.76	0.71	1.01	1.02
June-09	0.62	0.71	0.69	0.8	0.78
July-09	0.65	0.69	0.62	0.82	0.94
August-09	0.58	0.68	0.65	0.75	0.86
September-09	0.76	0.7	0.69	0.85	0.71
October-09	0.56	0.69	0.82	0.89	0.72
November-09	0.6	0.53	0.95	1.02	0.81
December-09	0.91	0.56	1.11	1.07	0.95
January-10	0.97	0.63	1.47	1.4	1.53
February-10	1.3	0.67	1.56	1.49	1.15
March-10	1.04	0.71	1.25	1.31	1.07
April-10	0.77	0.61	0.84	0.89	0.94
May-10	0.9	0.56	0.52	0.73	0.9
June-10	0.9	0.55	0.46	0.63	0.76
July-10	0.94	0.54	0.47	0.64	0.45
August-10	1.03	0.47	0.51	0.67	0.48
September-10	1.21	0.62	0.56	0.72	0.42

www.ingramcontent.com/pod-product-compliance
Lightning Source LLC
Chambersburg PA
CBHW081617170526
45166CB00009B/2998